Teach Yourself VISUALLY™

MacBook Pro and MacBook Air

7th Edition

by Guy Hart-Davis

Visual®

A Wiley Brand

Teach Yourself VISUALLY™ MacBook Pro and MacBook Air

7th Edition

Copyright © 2024 by John Wiley & Sons, Inc. All rights reserved.

Published by John Wiley & Sons, Inc., Hoboken, New Jersey.

Published simultaneously in Canada and the United Kingdom.

ISBNs: 9781394251322 (paperback), 9781394254385 (ePDF), 9781394254378 (ePub)

SKY10065022_011824

For general information on our other products and services or for technical support, please contact our Customer Care Department within the United States at (800) 762-2974, outside the United States at (317) 572-3993 or fax (317) 572-4002.

If you believe you've found a mistake in this book, please bring it to our attention by emailing our Reader Support team at wileysupport@wiley.com with the subject line "Possible Book Errata Submission."

Wiley also publishes its books in a variety of electronic formats. Some content that appears in print may not be available in electronic formats. For more information about Wiley products, visit our web site at www.wiley.com.

Library of Congress Cataloging in Publication data available on request.

Cover images: Laptop © SANALRENK/Getty Images; Screenshot Courtesy of Guy Hart-Davis

Cover design: Wiley

Sources: Apple Inc.: Chapter 3 opener, Figures 3.1 to 3.58, Chapter 5 opener, Figures 5.1 to 5.60, Chapter 6 opener, Figures 6.1 to 6.57, Chapter 7 opener, Figures 7.1 to 7.40, Chapter 8 opener, Figures 8.1 to 8.22, Chapter 10 opener, Figures 10.1 to 10.30, 10.32 to 10.36, and margin arts 600 to 644.

About the Author

Guy Hart-Davis is the author of more than 175 computer books, including *Killer ChatGPT Prompts: Harness the Power of AI for Success and Profit; iPhone For Dummies, 2024 Edition; macOS Sonoma For Dummies; Teach Yourself VISUALLY iPhone 14; Teach Yourself VISUALLY iPad; Teach Yourself VISUALLY Google Workspace; Teach Yourself VISUALLY Chromebook;* and *Teach Yourself VISUALLY Word 2019.*

Author's Acknowledgments

My thanks go to the many people who turned my manuscript into the highly graphical book you are holding. In particular, I thank Jim Minatel for asking me to write the book; Lynn Northrup for keeping me on track and skillfully editing the text; and Straive for laying out the book.

How to Use This Book

Who This Book Is For

This book is for the reader who has never used this particular technology or software application. It is also for readers who want to expand their knowledge.

The Conventions in This Book

① Steps

This book uses a step-by-step format to guide you easily through each task. **Numbered steps** are actions you must do; **bulleted steps** clarify a point, step, or optional feature; and **indented steps** give you the result.

② Notes

Notes give additional information — special conditions that may occur during an operation, a situation that you want to avoid, or a cross reference to a related area of the book.

③ Icons and Buttons

Icons and buttons show you exactly what you need to click to perform a step.

④ Tips

Tips offer additional information, including warnings and shortcuts.

⑤ Bold

Bold type shows command names, options, and text or numbers you must type.

⑥ Italics

Italic type introduces and defines a new term.

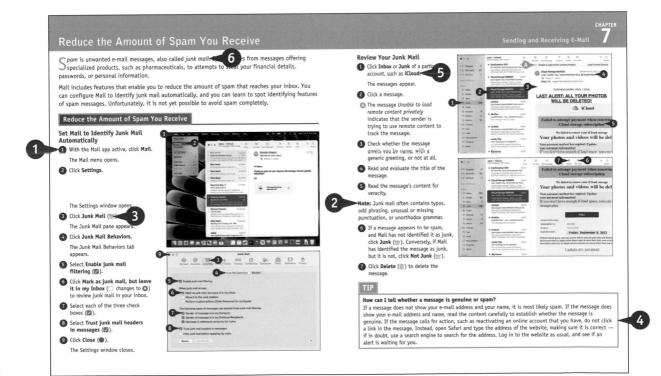

Table of Contents

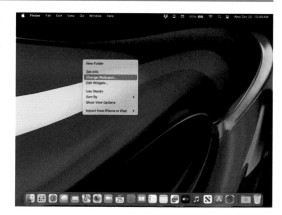

Chapter 3 Sharing Your MacBook with Others

Table of Contents

Table of Contents

Chapter 6 — Surfing the Web

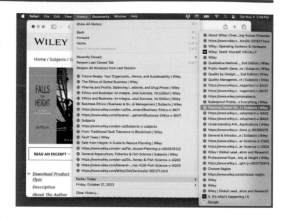

Chapter 7 — Sending and Receiving E-Mail

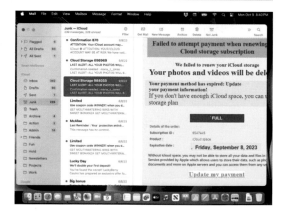

Chapter 8 — Chatting and Calling

Chapter 9 — Organizing Your Life

Table of Contents

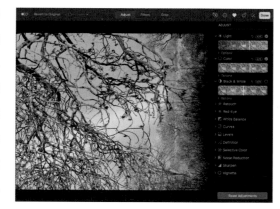

Chapter 12 Networking, Security, and Troubleshooting

CHAPTER 1

Getting Started with Your MacBook

Apple's MacBook laptops are among the best portable computers you can get. The powerful MacBook Pro and the lightweight MacBook Air enable you to work — or play — anywhere that suits you. Each MacBook comes with macOS, Apple's easy-to-use operating system. This chapter shows you how to set up your MacBook, navigate the macOS interface, and perform essential actions.

Understanding the MacBook Pro and MacBook Air

Macbook is the family name for Apple's laptop computers. As of this writing, the MacBook family consists of the powerful MacBook Pro models and the slim and lightweight MacBook Air models.

Each MacBook has similar core features, such as the display for viewing information and the keyboard and trackpad for entering data and controlling the computer. Beyond that, the MacBook models differ in various ways — from design, size, and weight to screen size, memory and storage capacity, and processor type and speed.

Identify Your MacBook's Main Features

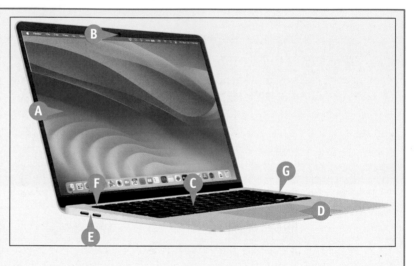

A Display
The MacBook's display provides a sharp, bright, and colorful view into all that you do.

B Camera
The built-in camera enables you to videoconference, take photos, and more.

C Keyboard
Along with the standard letter and number keys, the keyboard provides modifier keys — such as ⌘, Option, and Control — to control your MacBook. The keyboard has a backlight that illuminates automatically when you are using the MacBook in dim light, enabling you to see what you are doing.

D Trackpad
The trackpad enables you to manipulate objects on the screen using finger gestures. The entire trackpad is also the button that you click or double-click to give commands. On most MacBook models, you can also use a pressing movement called Force Touch to access commands quickly.

E USB-C Ports
The USB-C ports enable you to connect your MacBook to its power adapter and to other devices, such as external drives, external displays, iPhones, and iPads.

F Microphones
The microphones enable you to use your MacBook for audio and video calls without needing to connect a headset.

G Speakers
The speakers enable you to listen to music or other audio.

Identify the Ports on the MacBook Pro Models

Ⓐ Analog/Digital Audio In/Out

All the MacBook models include an analog/digital audio in/out port that enables you to connect an external microphone, headphones, or speakers. Beyond that, some MacBook Pro models feature MagSafe charging ports, an HDMI graphics port, and an SDXC card slot.

This port looks like a standard analog headphone port, but it works for both analog and digital audio and combines audio output and audio input. For analog audio output, simply connect headphones or analog speakers. For digital audio output, use a TOSLINK cable to connect digital audio equipment, such as surround-sound speakers. For audio input, connect a microphone or other sound input device.

Ⓑ MagSafe 3 Port

Connect the MacBook's power adapter to this port. The MagSafe 3 connector attaches magnetically, providing a secure connection but detaching easily if force is applied — for example, if someone's foot snags the power cord.

Ⓒ SDXC Card Slot

You can insert SDHC, SDXC, and other types of SD cards here so you can store files or transfer files to or from your MacBook.

The SDXC card slot accepts regular-size SD cards, which are 32mm × 24mm × 2.1mm. To use a miniSD card or a microSD card, get an adapter.

Standard-size SDXC cards protrude from the SDXC slot. This makes them easy to remove but even easier to damage if you leave them in the slot while transporting your MacBook. If you need to leave an SD card in the slot, get a microSD card and a low-profile adapter such as those made by BaseQi (www.baseqi.com).

Ⓓ USB-C Ports

The MacBook Pro models include two or three USB-C ports for connecting your USB devices. To connect a device that uses a cable with the flat, rectangular USB-A connector, you will need a USB-C-to-USB-A converter or a device that includes such a converter. If you need to connect multiple USB-A devices, consider getting a docking station that includes multiple USB-A ports.

Ⓔ HDMI Port

The 16-inch MacBook Pro includes one HDMI port for connecting an external display of up to 4K resolution or 8K resolution, depending on the model.

continued ▶

Understanding the MacBook Pro and MacBook Air (continued)

The current MacBook Air and MacBook Pro models include a row of hardware function keys above the keyboard. These keys are marked F1 through F12; each has a dedicated function, such as changing screen brightness or controlling media playback.

Some MacBook Pro models have the Touch Bar instead of the hardware function keys. The Touch Bar is a multitouch control strip whose contents change to suit the current app or selection. Some users find the Touch Bar convenient, but others prefer to have the physical function keys.

Meet Your MacBook's Keyboard

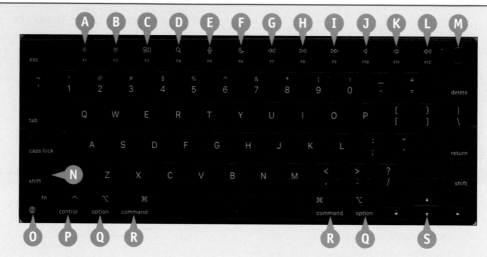

A Decrease Brightness

Press **F1** to decrease the screen's brightness.

B Increase Brightness

Press **F2** to increase the screen's brightness.

C Mission Control

Press **F3** to open Mission Control so you can quickly move between working spaces.

D Search

Press **F4** to open or close the Spotlight Search pane.

E Dictation

Press **F5** to start Dictation.

F Keyboard Backlight

Press **F6** to adjust the keyboard backlight.

G Previous/Rewind

Press **F7** to move to the previous item or rewind in Music and other applications.

H Play/Pause

Press **F8** to play or pause Music and other applications.

I Next/Fast-Forward

Press **F9** to move to the next item or fast-forward in Music and other applications.

J Mute

Press **F10** to mute your MacBook.

K Volume Down

Press **F11** to turn the volume down.

L Volume Up

Press **F12** to turn the volume up.

Ⓜ Power/Touch ID Button

Press the Power button to turn on your MacBook; press and hold the Power button to force your MacBook to turn off. For Touch ID, place your registered finger on the button without pressing.

Ⓝ Shift

Press `Shift` to type capital letters or the symbols that appear on the upper part of the keys.

Ⓞ Globe/Alternate Function

Hold down 🌐/`Fn` while pressing a function key to perform the alternate task.

Ⓟ Control

Press `Control` to give keyboard shortcuts.

Ⓠ Option

Press `Option` to give keyboard shortcuts.

Ⓡ Command

Press `Control` to give keyboard shortcuts.

Ⓢ Arrow Keys

Press ⬆, ⬇, ⬅, and ➡ to move the pointer around the screen.

Understanding and Using the Touch Bar

The Touch Bar is a flat sensor strip that replaces the row of physical function keys with virtual keys and controls that change depending on the app and the actions available to you. If your MacBook Pro has the Touch Bar, you can take a wide variety of actions from the Touch Bar by using the buttons and other controls that appear on it. This section shows you four examples of how the Touch Bar changes.

The Touch Bar can display keys for the dedicated functions of the hardware function keys:

When the Photos app is active, the Touch Bar can display controls for navigating among photos and performing common operations, such as marking a photo as a favorite, enhancing a photo, and rotating a photo:

Apps that enable you to manipulate color can display slider controls on the Touch Bar:

Blue: 116

When you need to perform calculations, the Touch Bar can display buttons for common operations:

Set Up Your MacBook

If you have just bought your MacBook, you need to set up macOS and create your user account before you can use it. Your user account is where you store your files and settings on the MacBook.

This section shows you the key decisions you make when setting up your MacBook. The first user account you create is an administrator account, which can create other accounts later for other users. You may also choose to create a personal account for yourself, leaving the administrator account strictly for administration.

Begin Setup and Choose Your Country

To begin setup, position your MacBook on a desk or table, connect its power supply, and then press the Power button. On most MacBook models, the Power button is at the upper-right corner of the keyboard and doubles as the fingerprint reader.

When the Language screen appears, click your language (A), and then click **Continue** (→, B). Then, on the Select Your Country or Region screen, select your country or region, and click **Continue** again.

Choose Written and Spoken Languages Settings

On the Written and Spoken Languages screen, verify that Preferred Languages (⊕, C) shows the language you want the macOS user interface to use, that Input Sources (⌨, D) shows the keyboard layout you want to use, and that Dictation (🎤, E) shows the language you will use for dictating text to your MacBook. If you want to change any of these settings, click **Customize Settings** (F) and then choose your preferred language, keyboard layout, or dictation language.

When the Written and Spoken Languages screen shows the settings you want, click **Continue** (G) to proceed.

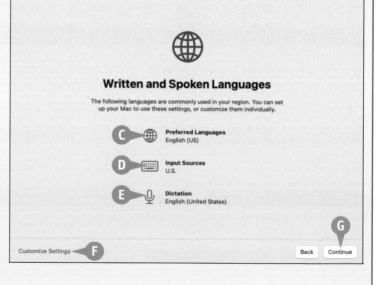

Apply Any Accessibility Settings You Need

On the Accessibility screen, you can choose whether to enable any of macOS's accessibility features now so that you can use them during setup and thereafter. Click **Vision** (👁, H) to enable features such as VoiceOver, Zoom, and Pointer Size. Click **Motor** (👆, I) to enable Accessibility Keyboard, an on-screen keyboard. Click **Hearing** (👂, J) to enable the Closed Captions feature and the Flash for Alerts feature. Click **Cognitive** (🧠, K) to enable features including Appearance, Speak Selection, and Typing Feedback.

You can enable and disable these accessibility features — and others — at any point after finishing setup. Click **Not Now** (L) if you do not want to set up any Accessibility features.

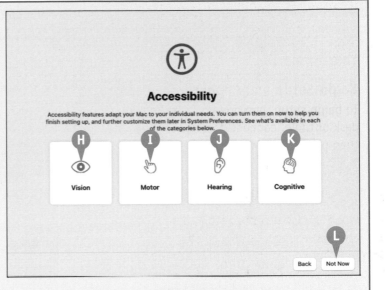

Choose Whether to Transfer Information to Your MacBook

On the Migration Assistant screen, you can choose whether to transfer information to your MacBook.

If you have information on another Mac, a Time Machine backup, or a Mac's startup disk, click **From a Mac, Time Machine backup or Startup disk** (○ changes to ◉, M), click **Continue**, and then follow the prompts.

If you have information on a Windows PC, click **From a Windows PC** (○ changes to ◉, N), click **Continue** (O), and then follow the prompts.

If you have no information to transfer, click **Not Now** (P).

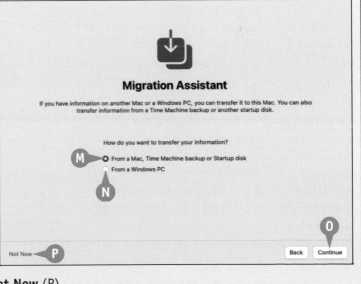

continued ▶

Set Up Your MacBook (continued)

When creating an account, you can use either your full name or a shortened version. You can edit the username that macOS suggests based on that name. You can choose whether to set a password hint to help yourself remember your password. You can also choose whether to let your Apple ID reset the password, enabling you to recover from a lost password by logging in using your Apple ID.

Sign In to Apple's Services with Your Apple ID

The Sign In with Your Apple ID screen enables you to sign in to Apple's services using your Apple ID, a credential consisting of an e-mail address and a password.

If you already have an Apple ID, type the e-mail address in the Apple ID box (A), and then click **Continue**. The Sign In with Your Apple ID screen then displays the Password box. Type your password in the Password box (B), and then click **Continue** (C).

If you do not have an Apple ID, you can click **Create new Apple ID** (D) and follow the prompts to create one.

If you prefer not to sign in with an Apple ID at this point, click **Set Up Later** (E).

Set Up Your Computer Account

On the Create a Computer Account screen, type your name the way you want it to appear in the Full Name box (F). In the Account Name box (G), macOS automatically enters a default account name consisting of your Full Name entry changed to lowercase and stripped of spaces and punctuation — for example, if you type *Maria Jones* as the full name, macOS suggests *mariajones* as the account name. You can edit the account name as needed.

Type a new password twice, once in each Password box (H). Optionally, click **Hint** (I) and type a password hint that will help you to recall your password.

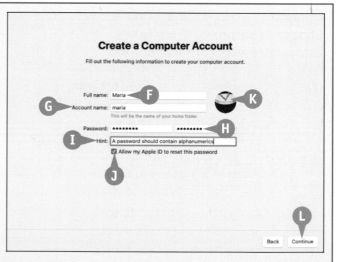

Select (✓) **Allow my Apple ID to reset this password** (J) if you want to be able to reset this password by using your Apple ID. This feature helps you avoid getting locked out of your MacBook.

Click the account icon (K) and choose the icon or image you want to use for your account. Then click **Continue** (L).

Choose Settings on the Make This Your New Mac Screen

The Make This Your New Mac screen summarizes the settings that the macOS installer will apply to your user account. Review the list of settings, which may include items such as Location Services, Device Analytics, App Analytics, Siri, Screen Time, and Appearance. If you are content to keep your existing settings, click **Continue**; otherwise, click **Customize Settings**, and then choose custom settings as needed.

Choose Whether to Use FileVault Disk Encryption

The FileVault Disk Encryption screen enables you to choose whether to use the FileVault feature to encrypt the data on your MacBook. Select **Turn on FileVault disk encryption** (☑) if you want to use FileVault; if not, deselect it (☐). If you select it, select (☑) or deselect (☐) **Allow my iCloud account to unlock my disk**; enabling this feature helps make sure you do not get locked out of your own data if you lose your FileVault password, but it means that anybody who compromises your iCloud account could also decrypt your FileVault data.

Enable Screen Time If You Need It

When the Screen Time screen appears, decide whether to activate the Screen Time feature, which tracks your or other users' computer usage so that you can analyze it. If you want to use Screen Time for either yourself or another user of your MacBook, click **Continue** (M); macOS enables the Screen Time feature as a whole, and you can configure the settings later. If you do not need Screen Time, click **Set Up Later** (N); should you want to use Screen Time later, you can enable it and configure it at that point.

After you finish configuring macOS, the desktop appears, and you can start using your MacBook as explained in the rest of this book.

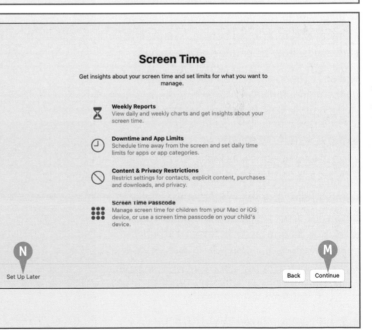

Start Your MacBook and Log In

When you are ready to start a computing session, start your MacBook and log in to macOS with the credentials for the user account you have set up or an administrator has created for you. After you start your MacBook, macOS loads and automatically displays the login screen by default or logs you in automatically. From the login screen, you can select your username and type your password.

When you log in, macOS displays the desktop with your apps and settings.

Start Your MacBook and Log In

 Press the Power button on your MacBook (not shown).

Note: This book uses "(not shown)" to indicate that a numbered step in text does not appear on the corresponding screen.

A screen showing the list of users appears.

Note: Your MacBook may not display the list of users on the login screen. Instead, it may simply log you in automatically or show a different login screen. Chapter 12 shows you how to change this behavior.

Note: If the login screen shows another user's name and icon, move the pointer over the name or icon to display the list of users.

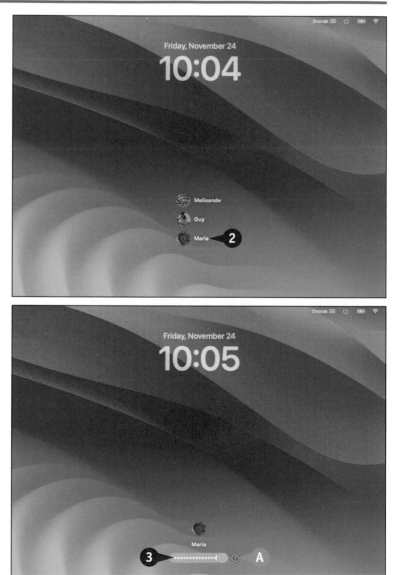

② Click your username.

Note: On a Touch ID–equipped MacBook, Touch ID is not always available for login. For example, you may need to type your password after restarting the MacBook, after not having used it for an extended time, or following multiple failed attempts to use Touch ID. This is a security measure.

The login window appears.

③ Type your password in the Enter Password box.

Ⓐ If you cannot remember your password, click **Hint** (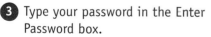).

B macOS displays your password hint.

C macOS also displays information about other actions you can take if you still cannot remember your password.

4 Type your password if you have not already done so.

5 Click **Log In** (⊙).

Note: Instead of clicking Log In (⊙), you can press **Return**.

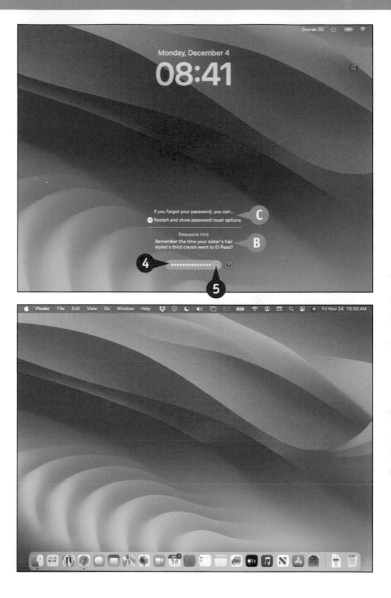

The MacBook displays your desktop, the menu bar, and the Dock. You can now start using the MacBook.

TIPS

Why does my MacBook go straight to the desktop instead of displaying the list of usernames?

Your MacBook is set to log in automatically. Logging in automatically is convenient when you are the only one who uses your MacBook, but it means that anyone who can start your MacBook can use it without providing credentials. Chapter 12 shows you how to turn off automatic login.

Why does my MacBook not show the list of usernames?

Hiding the list of usernames provides extra security and is widely used in companies, but it is usually not necessary for a MacBook used at home. Type your username in the Name field and your password in the Password field, and then click **Log In** (⊙).

Explore the macOS Desktop

Your MacBook runs the macOS operating system, which is currently in version 14, a version called Sonoma. The Macintosh operating system has long been known for being intuitive and is also pleasing to look at. It was the first major system interface to focus on graphical elements, such as icons. The macOS desktop is the overall window through which you view all that happens on your MacBook, such as looking at the contents of folders, working on documents, and surfing the Web.

Explore the macOS Desktop

Ⓐ Menu Bar

The menu bar usually appears at the top of the screen, showing the menus for the active application or app. macOS hides the menu bar in certain situations, such as when you display an app full screen.

Ⓑ Drives

The MacBook stores its data, including the software it needs to work, on an internal drive. This drive is a solid-state device, or SSD, rather than a traditional hard drive containing spinning platters, but it is often referred to as a "hard disk." You can also connect external drives for extra storage. You can choose whether to display an icon for each hard disk on the desktop.

Ⓒ iPod, iPhone, or iPad

You can connect one or more iPods, iPhones, or iPads to your MacBook to transfer files.

Ⓓ Folders

Folders are containers that you use to organize files and other folders stored on your MacBook.

Ⓔ Files

Files include documents, applications, or other sources of data. There are various kinds of documents, such as text, photos, graphics, songs, or movies.

Ⓕ Finder Windows

You view the contents of drives, folders, and other objects in Finder windows.

Ⓖ App and Document Windows

When you use apps, you use the windows that those apps display, for documents, web pages, games, and so on.

Work with the Finder Menu Bar and Menus

A **Apple Menu**

This menu is always visible so that you can access special commands, such as Shut Down and Log Out.

B **Finder Menu**

This menu enables you to control the Finder app itself. For example, you can display information about Finder or set preferences to control how it behaves.

C **File Menu**

This menu contains commands you can use to work with files and Finder windows.

D **Edit Menu**

This menu is not as useful in Finder as it is in other applications, but here you can undo what you have done or copy and paste information.

E **View Menu**

This menu enables you to determine how you view the desktop; it is especially useful for choosing Finder window views.

F **Go Menu**

This menu enables you to navigate to various places, such as specific folders.

G **Window Menu**

This menu enables you to navigate and arrange your open Finder windows.

H **Help Menu**

This menu provides help with macOS or the other applications.

I **Configurable Menus**

You can configure the menu bar to include specific menus, such as Screen Mirroring, Volume, Wi-Fi, Battery, and many more.

J **Fast User Switching**

This feature enables you to switch user accounts and open the Login window.

K **Spotlight Menu**

This menu enables you to search for information on your MacBook.

L **Control Center**

This pop-up panel gives you quick access to frequently used controls.

M **Clock**

Here you see the current day and time.

continued ▶

Explore the macOS Desktop (continued)

The Finder app controls the macOS desktop, and so you see the Finder menu bar whenever you work with the desktop. When you view the contents of a folder, you do so through a Finder window. There are many ways to view the contents of a Finder window, such as Icon view and List view. The sidebar enables you to quickly navigate the file system and to open files and folders with a single click. The Dock on the desktop and the sidebar in Finder windows enable you to access items quickly and easily.

Work with Finder Windows

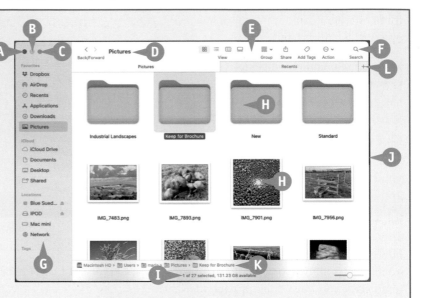

Ⓐ Close Button

Click to close a window.

Ⓑ Minimize Button

Click to shrink a window and move it onto the Dock.

Ⓒ Zoom Button

Click to expand a Finder window to the maximum size needed or possible; click it again to return to the previous size.

Ⓓ Window Title

The name of the location whose contents you see in the window.

Ⓔ Toolbar

Contains tools you use to work with files and folders.

Ⓕ Search Icon

Enables you display the Search box for finding files, folders, and other information.

Ⓖ Sidebar

Enables you to quickly access devices, folders, files, and tags, as well as searches you have saved.

Ⓗ Files and Folders

Shows the contents of a location within a window; this example shows the Icon view.

Ⓘ Status Bar

Shows information about the current location, such as the amount of free space when you are viewing the MacBook's drive.

Ⓙ Window Border

Drag a border or a corner to change the size of a window.

Ⓚ Path Bar

Shows the path to the location of the folder displayed in the window.

Ⓛ Tab Bar

Enables you to open multiple tabs containing different Finder locations within the same Finder window and quickly switch among them.

Work with the Dock and Sidebar

Ⓐ Favorites

Contains files, folders, searches, and other items that you can open by clicking them.

Ⓑ iCloud

Shows the folders you have stored in your space on iCloud Drive, such as Documents and Desktop.

Ⓒ Locations

Contains your MacBook's internal drive or drives, any DVD or CD in an external optical drive, external drives, network drives, and other devices that your MacBook can access.

Ⓓ Tags

Shows the list of tags you can apply to files and folders to help you identify and sort them easily.

Ⓔ Dock

Shows apps, files, and folders you can access with a single click, along with apps currently running.

Ⓕ Dock Divider Line

Divides the left side of the Dock from the right side. You can press `Control`+click the line to display the contextual menu for configuring the Dock.

Ⓖ Apps

Icons on the left side of the Dock are for apps; each open app has a dark dot under its icon unless you turn off this preference.

Ⓗ Files, Folders, and Minimized Windows

Icons on the right side of the Dock are for files, folders, and minimized windows. The default Dock includes the Downloads folder for files you download from the Internet.

Ⓘ Trash/Eject

macOS puts items you delete in the Trash; to get rid of them, you empty the Trash. When you select an ejectable device, such as a DVD, the Trash icon changes to the Eject icon.

Point and Click with the Trackpad

To tell the MacBook what you want to do, slide your finger across the trackpad to move the on-screen pointer over the object you want to work with. After you point to an object, you press the trackpad down to click, telling the computer what you want to do with the object. The number of times you click, and the manner in which you click, determine what happens to the object you point at.

Point and Click with the Trackpad

Point and Click

1 Slide your finger across the trackpad until the pointer points at the appropriate icon.

2 Press the trackpad once to click the trackpad (not shown). This is a single click.

A The object becomes highlighted, indicating that it is now selected.

Double-Click

1 Slide your finger across the trackpad until the pointer points at the appropriate icon (not shown).

2 Click the trackpad twice (not shown).

Your selection opens.

Point, Click, and Drag

1 Slide your finger across the trackpad until the pointer points at the appropriate icon.

2 Press down the trackpad and hold it (not shown).

The object that you were pointing at becomes attached to the arrow and remains so until you release the trackpad.

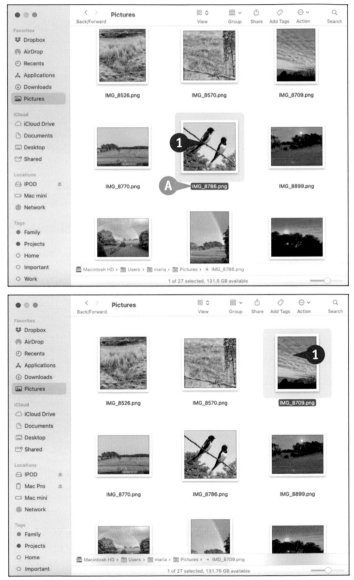

3 Drag your finger on the trackpad to move the object.

4 When you get to the object's new position, release the trackpad.

Note: Dragging an item to a different drive, flash drive, or disk volume copies it there. Changing an item's location on the same drive moves the item instead.

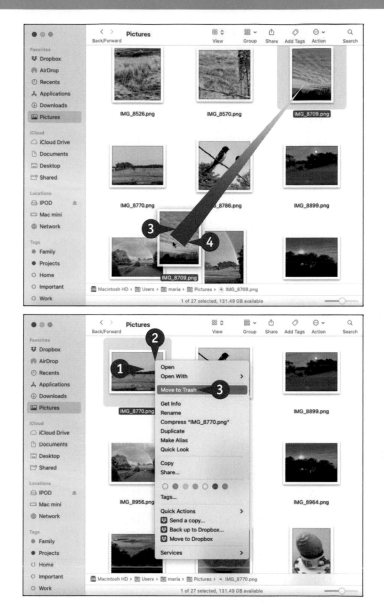

Secondary Click (Control+Click)

1 Point to an object in a Finder window or on the desktop, or to the desktop itself.

Note: To select more than one item at the same time, click the first item, and then press and hold while you click each other item.

2 Press (Control)+click the trackpad.

A contextual menu appears.

3 Point to the appropriate command on the menu and click the trackpad once to give the command.

TIP

Why do things I click stick to the arrow?

You can configure the trackpad so you can drag items without having to hold down the trackpad. When this setting is on and you click an item, it gets attached to the pointer. When you move the pointer, the item moves, too. To configure this setting, see the section "Configure the Trackpad or Other Pointing Device" in Chapter 2.

Connect to a Wireless Network

If a wireless network is available, you can connect your MacBook to it. Wireless networks are convenient for both homes and businesses because they require no cables and are fast and easy to set up.

Your MacBook includes a wireless network feature that uses some of the wireless network standards called Wi-Fi. You can control wireless networks directly from the Wi-Fi menu at the right end of the menu bar. To connect to a Wi-Fi network, you need to know its name and password.

Connect to a Wireless Network

Note: If you connected your MacBook to a wireless network during setup, you do not need to set up the connection to the same network again.

1️⃣ Click **Wi-Fi status** (🛜) on the menu bar.

The menu opens.

Note: If the list of wireless networks appears on the menu, go to step **4**.

2️⃣ Set the **Wi-Fi** switch to On (⬤).

macOS turns Wi-Fi on.

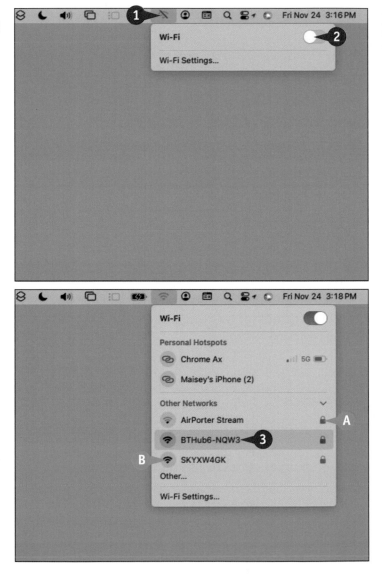

Note: If the Wi-Fi menu closes, click **Wi-Fi status** (🛜) on the menu bar to reopen it.

The Wi-Fi menu displays a list of the wireless networks your MacBook can detect.

🅐 A lock icon (🔒) indicates that the network is secured with a password or other security mechanism.

🅑 The signal strength icon (🛜) indicates the relative strength of the network's signal.

3️⃣ Click the network to which you want to connect your MacBook.

If the wireless network uses a password, your MacBook prompts you to enter it.

4 Type the password in the Password box.

C If you want to see the characters of the password to help you type it, click **Show password** (changes to).

5 Click **Join**.

Your MacBook connects to the wireless network, and you can start using network resources.

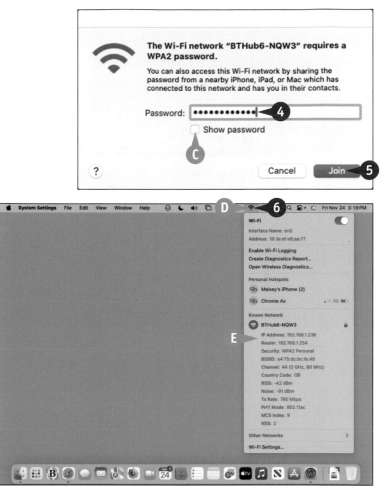

D The number of arcs on the Wi-Fi status icon ( ) indicates the strength of the connection, and ranges from one arc to four arcs.

6 To see more details about the wireless network, press Option+click **Wi-Fi status** ( ) on the menu bar.

E The network's details appear, including the physical mode, the wireless channel, and the security type.

TIPS

How do I disconnect from a wireless network?

When you have finished using a wireless network, you can disconnect from it by turning Wi-Fi off. Click **Wi-Fi status** ( ) on the menu bar and then set the **Wi-Fi** switch to Off ( ).

What kind of Wi-Fi network do I need for my MacBook?

Wi-Fi networks use several different standards. As of this writing, the latest standard that MacBook models support is 802.11ax, which provides very fast data rates. Your MacBook can also use older Wi-Fi standards, such as 802.11a, 802.11b, 802.11g, and 802.11n, so you can use most Wi-Fi networks. If your wireless router is several years old, look into upgrading to a newer model, as it may provide a substantial increase in speed.

Give Commands

The easiest ways to give commands in macOS are by using the menus and the toolbar. You can also give commands by pressing keyboard shortcuts.

The menu bar at the top of the window shows the Apple menu (🍎) on the left followed by the menus for the active app. Any open window can have a toolbar, usually across its top but sometimes elsewhere in the window.

Give Commands

Give a Command from a Menu

① On the Dock, click the app you want to activate — **Finder** (😀) in this example.

Note: You can also click the app's window if you can see it.

② On the menu bar, click the menu you want to open.

The menu opens.

③ Click the command you want to give.

The app performs the action associated with the command.

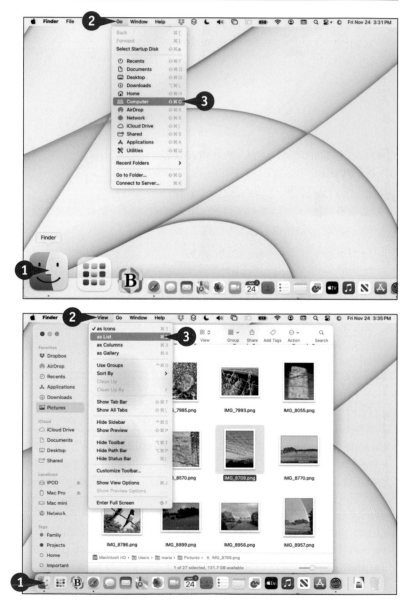

Choose Among Groups of Features on a Menu

① On the Dock, click the app you want to activate — **Finder** (😀) in this example.

② On the menu bar, click the menu you want to open.

The app opens the menu.

③ Click the option you want to use.

The app activates the feature you selected.

Give a Command from a Toolbar

1 On the Dock, click the app you want to activate — **Finder** (🙂) in this example.

2 Click the button for the command on the toolbar, or click a pop-up menu and then click the menu item for the command.

The app performs the action associated with the toolbar button or menu item.

Choose Among Groups of Features on a Toolbar

1 On the Dock, click the app you want to activate — **Finder** (🙂) in this example.

2 In the group of buttons, click the button you want to choose.

Ⓐ The app highlights the button you clicked to indicate that the feature is turned on.

Ⓑ The app removes highlighting from the button that was previously selected.

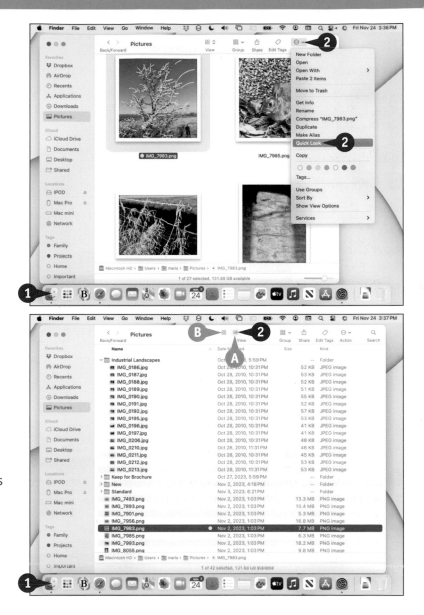

TIP

Is it better to use the menus or the toolbar?

If the toolbar contains the command you need, using the toolbar is usually faster and easier than using the menus. You can customize the toolbar in many apps by opening the **View** menu, choosing **Customize Toolbar**, working in the dialog that opens, and then clicking **Done.** Use this command, or other similar commands, to place the buttons for your most-used commands just a click away.

Open, Close, and Manage Windows

M ost macOS apps use windows to display information so that you can see it and work with it. You can resize most windows to the size you need or expand a window so that it fills the screen. You can move windows and position them so that you can see those windows you require, minimize other windows to icons on the Dock, or hide an app's windows from view.

Open, Close, and Manage Windows

Open a Window

1 Click anywhere on the desktop.

macOS activates Finder and displays the menu bar for it.

Note: Clicking anywhere on the desktop activates Finder because the desktop is a special Finder window. You can also click **Finder** (🙂) on the Dock.

2 Click **File**.

The File menu opens.

3 Click **New Finder Window**.

A Finder window opens, showing your files in your default view.

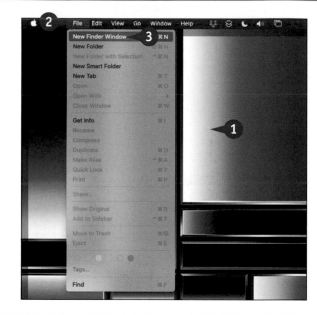

Move, Resize, and Zoom a Window

1 Click the window's title bar and drag the window to where you want it.

2 Click a border or corner of the window and drag until the window is the size and shape you want.

3 Click **Zoom** (🟢).

The window zooms to full screen.

4 Move the pointer to the upper-left corner of the screen.

The macOS menu bar and the app's title bar appear.

5 Click **Zoom Back** (🟡; not shown).

The window zooms back to its previous size.

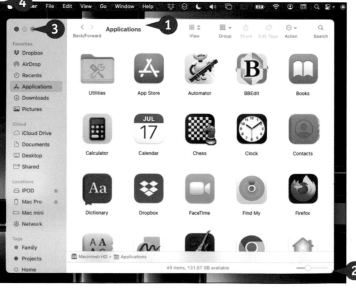

Close a Window

1 Click **Close** (●).

Note: When you move the pointer over the upper-left corner of a window, Close (●) changes to Close (✕), Minimize (●) changes to Minimize (⊖), and Zoom (●) changes to Zoom (⊘).

The window closes.

Note: You can also close a window by pressing ⌘+W. To close all the windows of the app, press Option+click **Close** (●) or press ⌘+Option+W.

Minimize or Hide a Window

1 Click **Minimize** (●).

macOS minimizes the window to an icon on the right side of the Dock.

Note: You can also minimize a window by pressing ⌘+M.

2 Click the icon for the minimized window.

macOS restores the window to its original size and position.

TIP

How can I find out where a document in a window is located?
To quickly see what folder contains a file or folder, press ⌘+click the window's name in the title bar. The window displays a pop-up menu showing the path of folders to this folder. Click a folder in the path to display that folder in Finder, or click the title bar to hide the pop-up menu again.

Using Control Center

macOS's Control Center feature gives you quick access to a range of frequently needed controls. These controls include Wi-Fi, Bluetooth, and AirDrop; Focus, Stage Manager, and Screen Mirroring; Display brightness, Sound volume and AirPlay; and the Now Playing item in the Music app.

You open Control Center by clicking its icon toward the right end of the menu bar. Control Center opens as a pane on the right side of the screen, showing the top layer of controls. You can display further layers to reach other controls.

Using Control Center

Open Control Center and Toggle Settings On and Off

1 Click **Control Center** (🎛) to open Control Center.

2 Click **Wi-Fi** (🛜 changes to ⌢) to turn Wi-Fi off.

3 Click **Bluetooth** (🔵 changes to ✳) to turn Bluetooth off.

4 Click **AirDrop** (🔵 changes to 🔘) to turn AirDrop off.

5 Click **Do Not Disturb** (🌙 changes to 🌙) to turn Do Not Disturb on.

6 Click outside Control Center to close Control Center.

Change Display Brightness and Configure Display Settings

1 Click **Control Center** (🎛) to open Control Center.

2 Drag the **Brightness** slider to set the screen brightness.

3 Move the pointer over the Display area.

Ⓐ An Expand button (>) appears.

4 Click either **Expand** (>) or anywhere on the Display heading to open the Display panel.

The Display panel opens.

5 Click **Dark Mode** (◐ changes to ◉) to turn Dark Mode on.

6 Click **Night Shift** (☀ changes to ☀) to turn Night Shift on.

7 Click **True Tone** (☀ changes to ☀) to turn True Tone on.

8 Click outside the Display panel to close the panel.

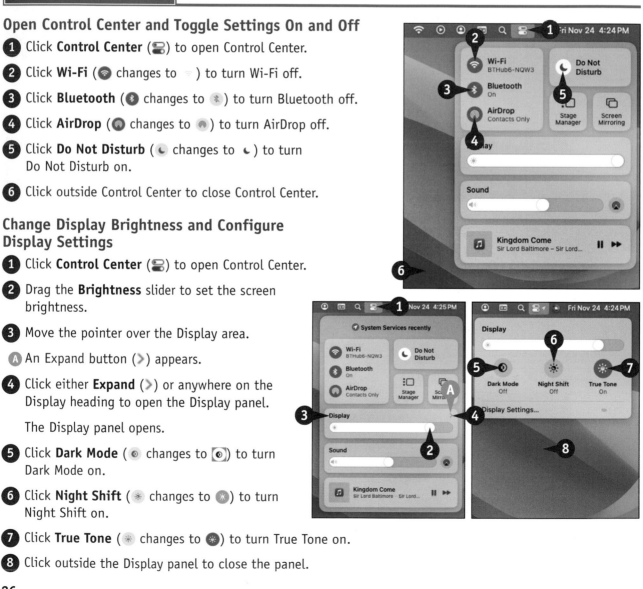

26

Change the Sound Volume and Select an Output Device

1 Click **Control Center** () to open Control Center.

2 Drag the **Sound** slider to set the sound volume.

3 Click the **Sound** heading to open the Sound panel.

4 In the Output list, click the output device you want to use.

Your MacBook directs the sound output to the device you selected.

5 Click outside the Sound panel to close the Sound panel.

Apply a Focus

1 Click **Control Center** () to open Control Center.

2 Click the **Do Not Disturb** heading to open the Focus panel.

Note: Click any part of the Do Not Disturb button except for the icon ().

3 Click the focus you want to apply.

The focus becomes active.

4 Click outside the Focus panel to close the Focus panel.

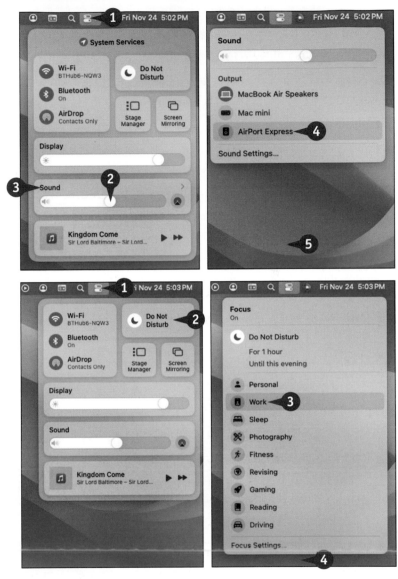

TIP

Can I change the controls displayed in Control Center?
Yes, macOS enables you to customize Control Center to some extent. See the section "Configure Control Center and the Menu Bar" in Chapter 2 for details.

Using Notifications

macOS's Notification Center feature keeps you up to date with what is happening in your apps. Notification Center puts all your alerts, from incoming e-mail messages and instant messages to calendar requests and software updates, in a single place where you can easily access and manage them.

You open Notification Center by clicking the clock readout at the right end of the menu bar. Notification Center opens as a pane on the right side of the screen, and it contains sections you can expand or collapse as needed.

Using Notifications

View a Notification

Ⓐ When you receive a notification, a notification banner appears in the upper-right corner of the screen for a few seconds.

Note: Notification Center can display either banners or alerts. A banner appears for a few seconds and then disappears. An alert remains on-screen until you dismiss it.

① If you want to see the item that produced the notification, click the banner.

Ⓑ To see if there are other actions you can take with the notification, move the pointer over it.

Ⓒ If the Options button appears, click **Options**.

The Options menu opens.

Ⓓ You can then click an action, such as Reply.

Display Notification Center When Your Desktop Is Visible

① Click the clock readout.

Notification Center opens.

Ⓔ You can dismiss a notification by moving the pointer over it and then clicking **Close** (×).

② Optionally, click a notification to display the related item in its app.

③ When you are ready to close Notification Center, click outside it.

Note: You can also click the clock readout to close Notification Center.

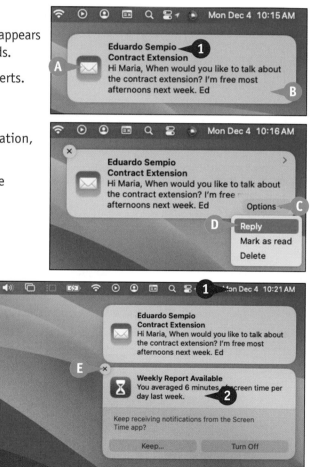

Choose What Notifications to Display and Configure Focus Settings

1 Click **System Settings** (⚙) on the Dock.

2 Click **Notifications** (🔔) in the sidebar.

3 Click the **Show Previews** pop-up menu (↕) and click **Always**, **When Unlocked**, or **Never**.

4 Set the **Allow notifications when the display is sleeping** switch to On (⬤) or Off (◯).

5 Set the **Allow notifications when the screen is locked** switch to On (⬤) or Off (◯).

6 Set the **Allow notifications when mirroring or sharing the display** switch to On (⬤) or Off (◯).

7 In the Application Notifications area, click each app in turn and set notification options.

8 Click **Focus** (🌙) in the sidebar.

9 Click each focus you want to configure, and then choose settings.

10 Set the **Share across devices** switch to On (⬤) if you want all your iCloud devices to use these Focus settings.

11 Click **Focus status** and choose settings. See the tip for advice.

12 Click **Close** (⬤) to close System Settings.

TIP

What is Focus status?

The Focus status settings enable you to specify whether apps can tell other people that you're using a particular focus and have notifications silenced. Click **Focus status** at the bottom of the Focus pane in System Settings to display the Focus Status pane, and then set the **Share Focus status** switch to On (⬤) on Off (◯), as needed. If you set the switch to On (⬤), go to the **Share From** list and set the switch for each focus to On (⬤) or Off (◯), as appropriate. For example, you might let apps tell people when your Work Focus is enabled but not when your Gaming Focus is enabled.

Add Widgets to Your Desktop

A *widget* is a miniature app that displays a particular kind of information. For example, the Weather widget tells you about the current weather, the Reminders widget displays upcoming and overdue tasks, and the Stocks widget shows you how your stocks are soaring or sinking.

Earlier version of macOS kept widgets in Notification Center, but macOS Sonoma enables you to put widgets directly on the desktop. You can set macOS to display the widgets all the time, when you are using the Stage Manager app-management feature, or both.

Add Widgets to Your Desktop

Add and Remove Widgets

1 Press **Control**+click open space on the desktop.

The contextual menu opens.

2 Click **Edit Widgets**.

Ⓐ The Widgets dialog opens.

Ⓑ The All Widgets category in the sidebar is selected at first.

Ⓒ Frequently used widgets appear.

3 To add a widget, move the pointer over it, and then click **Add** (➕).

Ⓓ macOS adds the widget in the first available position.

Ⓔ You can click **Search Widgets** (🔍) and search by keyword.

4 Click a category. This example uses **Clock** (🕐).

The widgets in that category appear.

5 Scroll down if necessary to view other widgets.

6 Drag a widget you want to where you want it on the desktop.

Ⓕ You can click **Remove** (➖) to remove a widget.

7 When you finish adding widgets, click **Done**.

The Widgets dialog closes.

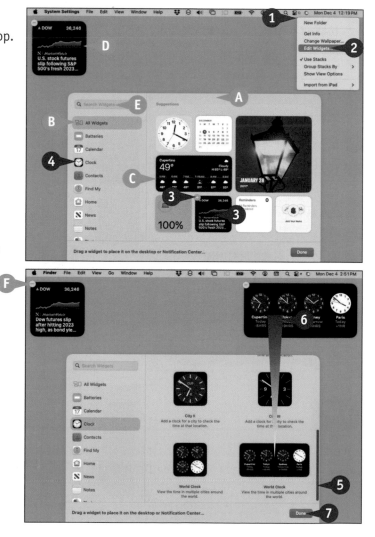

Edit a Widget

Note: To move a widget to a different location, drag it to where you want it.

1 Press `Control`+click the widget.

The contextual menu opens.

G If the Size section appears on the contextual menu, you can click a different size to switch the widget to that size.

2 Click the Edit item, such as **Edit "Clock"** in this example.

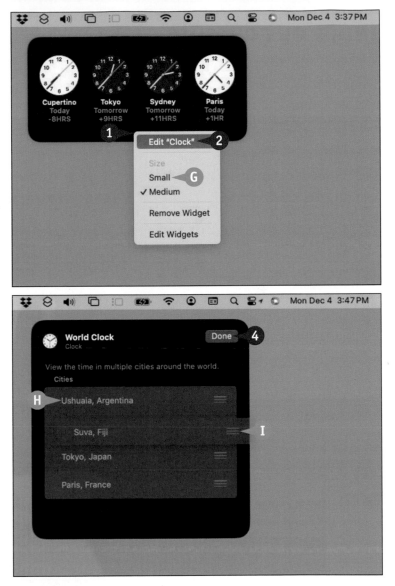

The widget opens for editing.

3 Use the controls displayed to edit the widget. For example, for the World Clock widget, you can change the cities displayed and the order in which they appear:

H You can click a city to display the Cities panel, search or browse for the city you want, and then click it.

I You can drag a city up or down by its handle (▤).

4 Click **Done**.

The widget reappears with its new settings.

TIP

How do I control when macOS displays my desktop widgets?
Click **System Settings** (⚙) on the Dock to open the System Settings window, and then click **Desktop & Dock** (▣). In the Desktop & Dock pane, scroll down to the Widgets section. Select (☑) or clear (☐) the **On Desktop** check box to control whether the widgets appear on the desktop. Select (☑) or clear (☐) the **In Stage Manager** check box to control whether the widgets appear when Stage Manager is active. Click **Close** (●) to close the System Settings window.

Put Your MacBook to Sleep and Wake It Up

m acOS enables you to put your MacBook to sleep easily and wake it up quickly. So when you are ready for a break but you do not want to end your computing session, put the MacBook to sleep instead of shutting it down.

Sleep uses only a minimal amount of power but keeps all your apps open and lets you start computing again quickly. When you wake your MacBook up, your apps and windows are where you left them, so you can swiftly resume what you were doing.

Put Your MacBook to Sleep and Wake It Up

Put Your MacBook to Sleep

1 Click **Apple** (🍎).

The Apple menu opens.

Note: You can also put your MacBook to sleep by closing its lid.

2 Click **Sleep**.

The MacBook turns its screen off and puts itself to sleep.

Note: You can also put your MacBook to sleep by pressing its Power button for a moment.

Wake Your MacBook

1 Click the trackpad or press any key on the keyboard, or rest your finger on the Touch ID sensor.

Note: If you put your MacBook to sleep by closing its lid, lift the lid instead.

Your MacBook wakes up and turns on the screen. All the apps and windows that you were using are open where you left them.

Your MacBook reestablishes any network connections that it normally uses and performs regular tasks, such as checking for new e-mail.

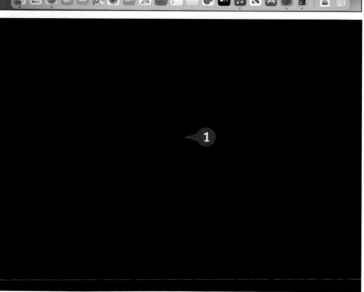

Log Out, Shut Down, and Resume

When you have finished using your MacBook for now, end your computing session by logging out. From the login screen, you can log back in when you are ready to use your MacBook again. When you have finished using your MacBook and plan to leave it several days, shut it down.

Whether you log out or shut down your MacBook, you can choose whether to have macOS reopen your apps and documents when you log back on. This helpful feature can help you get back to work — or play — quickly and easily.

Log Out, Shut Down, and Resume

Log Out from Your MacBook

1 Click **Apple** (🍎).

The Apple menu opens.

2 Click **Log Out**.

A dialog opens asking if you are sure you want to log out.

3 Click **Reopen windows when logging back in** (⬜ changes to ✅) if you want to resume your apps and documents.

4 Click **Log Out**.

Your MacBook displays the window showing the list of users. You or another user can click your name to start logging in.

Shut Down Your MacBook

1 Click **Apple** (🍎).

The Apple menu opens.

Ⓐ You can click **Restart** to restart your MacBook instead of shutting it down.

2 Click **Shut Down**.

A dialog opens asking if you are sure you want to shut down.

3 Click **Reopen windows when logging back in** (⬜ changes to ✅) if you want to resume your apps and documents.

4 Click **Shut Down**.

The screen goes blank, and your MacBook switches itself off.

CHAPTER 2

Configuring Your MacBook

You can customize many aspects of macOS to make it work the way you prefer. You can change the wallpaper; personalize the Dock icons, the menu bar, and Control Center; and adjust the keyboard and trackpad or other pointing device. You can also run apps or open specific documents each time you log in or set your MacBook to go to sleep automatically when you are not using it.

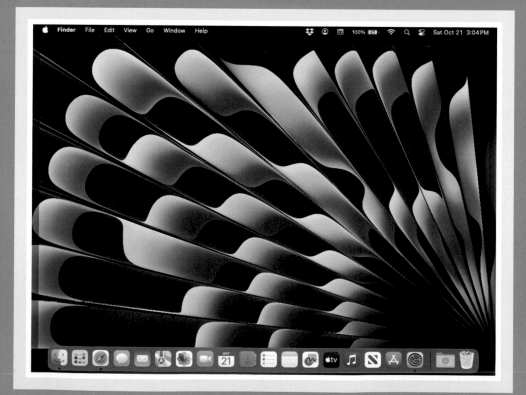

Change the Wallpaper

macOS enables you to change the wallpaper, the desktop background, to show the picture you prefer. macOS includes many varied desktop pictures and solid colors, but you can also set any of your own photos as the wallpaper. You can tile, stretch, or crop the photo to fill the screen or center it on the screen.

You can also choose between displaying a single picture on the desktop and displaying a series of images that change automatically.

Change the Wallpaper

Display the Wallpaper Pane in System Settings

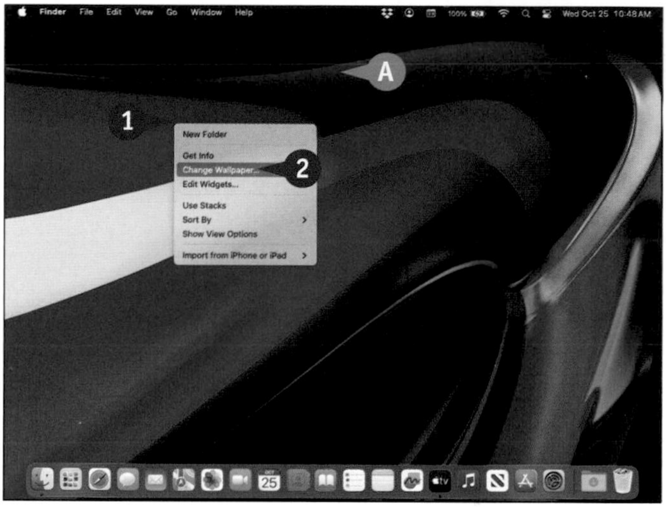

Ⓐ The wallpaper is visible when no window is open in front of it.

Note: A *dynamic* wallpaper can automatically change its colors to match the light conditions in your MacBook's location.

1. Press Control + click the desktop.

   The contextual menu opens.

2. Click **Change Wallpaper**.

   The Wallpaper pane in System Settings appears.

   You can now apply a built-in wallpaper, one or more photos, or a color, as explained in the following subsections.

Apply a Built-In Wallpaper

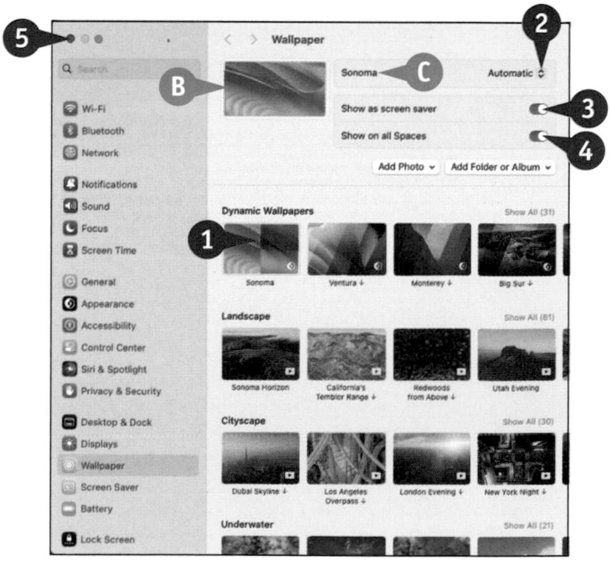

1. Click the wallpaper you want to apply.

Ⓑ The wallpaper's preview appears.

Ⓒ The wallpaper's name appears.

2. For a dynamic wallpaper, click ⌃⌄, and then click **Automatic**, **Light**, or **Dark**, as needed.
3. Set the **Show as screen saver** switch to On (●) or Off (○), as needed.
4. Set the **Show on all Spaces** switch to On (●) or Off (○), as needed.
5. Click **Close** (●).

   The System Settings window closes.

Apply One or More Photos as Wallpaper

1 Click **Add Photo** ( ).

D You can click **Add Folder or Album** ( ) to add a folder or album of photos.

2 Click **From Photos** to display the Photos browser, and then click the photo to use.

E You can click **Choose** to select a photo from your MacBook's file system.

F A preview of the photo appears.

G The wallpaper shows the photo.

H You can click **Cycle** ( ) to cycle through the last few photos you chose.

I You can click **Remove** ( ) to remove a photo.

3 Click and click the way to fit the photo to the screen. See the tip for details.

4 Click **Close** ( ).

The System Settings window closes.

Apply a Color as Wallpaper

1 Scroll down to the bottom of the Wallpaper pane.

J You can click **Show All** to display all available colors.

K You can click **Cycle** ( ) to cycle through all the colors.

L You can click Add (+) and use the Colors window to create a custom color.

2 Click the color.

M The color appears.

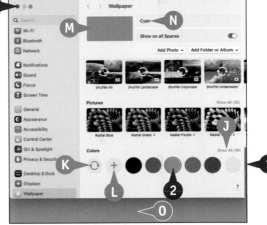

N The color name appears.

O The wallpaper shows the color.

3 Click **Close** ( ).

The System Settings window closes.

TIP

Which option should I choose for fitting the image to the screen?

In the Desktop & Screen Saver pane, choose **Fit to Screen** to match the image's height or width — whichever is nearest — to the screen. Choose **Fill Screen** to make an image fill the screen without distortion but cropping off parts that do not fit. Choose **Stretch to Fill Screen** to stretch the image to fit the screen exactly, distorting it as needed. Choose **Center** to display the image at full size in the middle of the desktop. Choose **Tile** to cover the desktop with multiple copies of the image.

Set Up a Screen Saver

macOS enables you to set a screen saver to hide what your screen is showing when you leave your MacBook idle. A *screen saver* is an image, a sequence of images, a moving pattern, or a video that appears on the screen. You can choose what screen saver to use and the delay before it starts. If you apply one of macOS' moving wallpapers, you can use that wallpaper as the screen saver, too.

macOS comes with a variety of attractive screen savers. You can download other screen savers from websites.

Set Up a Screen Saver

1 Press **Control**+click the desktop.

The contextual menu opens.

2 Click **Change Wallpaper**.

The Wallpaper pane of System Settings appears.

A If your MacBook is using a moving wallpaper, and you want to use that wallpaper as the screen saver as well, set the **Show as screen saver** switch to On (⊂●) and go to step **11**.

3 Click **Screen Saver** (▣).

The Screen Saver pane appears.

B You can click **Show All** to display all the screen savers in a category.

4 Click the screen saver you want to use.

C The screen saver's preview appears.

D The screen saver's name appears.

5 Set the **Show as wallpaper** switch to On (⊂●) if you want to use the screen saver as wallpaper.

6 Set the **Show on all Spaces** switch to On (⊂●) if you want to use the screen saver on all Spaces.

7 Click the preview.

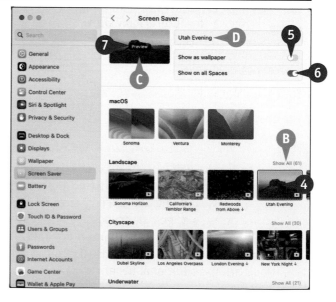

The screen saver preview appears full screen.

8 Click anywhere on the screen saver when you want to stop the preview.

The Screen Saver pane appears again.

Note: Video screen savers take up storage on your MacBook and consume more power to display than other types of screen savers.

9 Click **Lock Screen**.

The Lock Screen screen appears.

10 Click **Start Screen Saver when inactive** (⌄), and then click the delay to use. Your option run from **For 1 minute** to **For 3 hours** and **Never**.

Note: The *Display will sleep before screen saver starts* warning appears if the Start Screen Saver When Inactive time is equal to or greater than both the Turn Display Off on Battery When Inactive setting and the Turn Display Off on Power Adapter When Inactive setting.

11 Click **Close** ().

System Settings closes.

TIP

Must I use a screen saver to protect my MacBook's screen from damage?

No. Screen savers originally protected cathode ray tube — CRT — displays from having static images "burned in" to their screens. LCD and LED screens, such as that on your MacBook, do not suffer from this problem, so you need not use a screen saver; however, future screens based on technologies such as OLED — organic light-emitting diode — may suffer burn-in problems. Nowadays you can use a screen saver to protect the information on-screen or to provide visual entertainment — but the best move is to put your MacBook to sleep, which has the additional benefit of saving battery power.

Configure Battery and Sleep Settings

The Battery settings pane enables you to view your MacBook's battery level and battery usage, configure power settings separately for when your MacBook is running from the battery and from the power adapter, and set up a schedule for waking your MacBook or making it sleep. You can set your MacBook to turn off the display after a period of inactivity, dim the display on battery power, and enable Low Power Mode. You can put your MacBook to sleep manually at any time by closing the lid or by clicking **Apple** (**) and **Sleep**.

Configure Battery and Sleep Settings

1 Click **System Settings** () on the Dock.

The System Settings window opens.

2 Click **Battery** ().

The Battery pane appears.

3 If you want to configure Low Power Mode, click the **Low Power Mode** pop-up menu, and then click **Never**, **Always**, **Only on Battery**, or **Only on Power Adapter**, as needed.

4 Click **Battery Health** ().

The Battery Health dialog opens.

5 Verify that the Battery Condition readout says *Normal*. If it says *Service Recommended*, contact Apple or an Apple Authorized Service Provider for advice.

A The Maximum Capacity readout shows how much of the battery's design capacity remains, such as 83% remaining instead of 100%.

6 Set the **Optimized Battery Charging** switch to On () if you want your MacBook to develop a customized charging routine based on your detected usage.

7 Click **Done**.

The Battery Health dialog closes.

8 Click **Last 24 Hours**.

B The *Last Charged to 100%* readout shows when the battery last reached 100% charge.

C The Battery Level histogram shows the battery levels for the last 24 hours.

D The green horizontal line shows when your MacBook was connected to a power adapter.

E The Screen On Usage histogram shows the amount of time the screen was on each hour for the last 24 hours.

9 Click **Last 10 Days** and examine the Energy Usage histogram and the Screen On Usage histogram for the last 10 days.

10 Click **Options**.

The Options dialog opens.

11 Set the **Slightly dim the display on battery** switch to On (⬤) to reduce power usage by dimming the display.

12 Set the **Prevent automatic sleeping on power adapter when the display is off** switch to On (⬤) to prevent your MacBook from going to sleep when the power adapter is plugged in.

13 Click the **Wake for network access** pop-up menu (⬍), and then click **Always**, **Only on Power Adapter**, or **Never**, as needed.

14 Set the **Optimize video streaming while on battery** switch to On (⬤) to save battery power at the expense of video brightness and color quality.

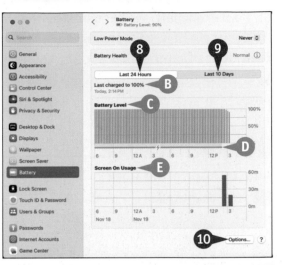

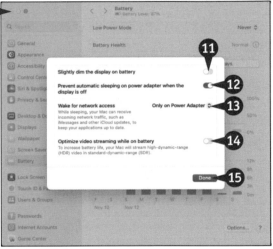

15 Click **Done**.

The Options dialog closes.

16 Click **Close** (⬤).

System Settings closes.

TIP

Which button do I press to wake my MacBook from sleep?
You can press any key. If you are not certain whether the MacBook is asleep or preparing to run a screen saver, press `Shift`, `Control`, `Option`, or `⌘`. These keys do not type a character if the MacBook turns out to be awake instead of asleep. You can also move your finger on the trackpad to wake your MacBook.

Customize the Dock

macOS enables you to customize the Dock so that it contains the icons you find most useful and it appears in your preferred position on the screen. You can add apps, files, or folders to the Dock; reposition the Dock's icons; and remove most of the existing items if you do not need them.

To customize the Dock, you drag items to it, from it, or along it. You can also use the Dock's contextual menu to change the Dock's position, configuration, or behavior.

Customize the Dock

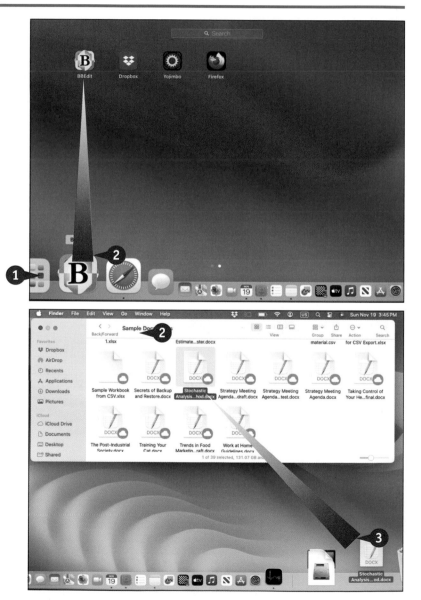

Add an App to the Dock

1 Click **Launchpad** (▦) on the Dock.

The Launchpad screen appears.

2 Drag the app to the left side of the divider line on the Dock.

The app's icon appears on the Dock.

Note: You can also add an app to the Dock by opening the app, pressing `Control` + clicking its Dock icon, highlighting or clicking **Options**, and then clicking **Keep in Dock**.

Add a File or Folder to the Dock

1 Click **Finder** (🙂) on the Dock.

2 In the Finder window, navigate to the file or folder you want to add to the Dock.

3 Drag the file or folder to the right side of the divider line on the Dock.

The item's icon appears on the Dock.

Note: When you drag a file or folder to the Dock, macOS creates a link to the file or folder rather than moving the original item.

Remove an Item from the Dock

1 If the app is running, press Control+click its Dock icon and then click **Quit** on the contextual menu (not shown).

2 Press Control+click the app's icon on the Dock.

The contextual menu opens.

3 Highlight or click **Options**.

The Options continuation menu opens.

4 Click **Remove from Dock**.

macOS removes the icon from the Dock.

Note: You can also click an icon and drag it from the Dock toward the middle of the desktop. When a Remove pop-up message appears, release the icon.

Configure the Dock

1 Press Control+click the Dock divider bar.

Ⓐ Click **Turn Hiding On** to hide the Dock when the pointer is not over it.

Ⓑ Click **Turn Magnification Off** to turn off magnification.

Ⓒ Click **Position on Screen** and then click **Left**, **Bottom**, or **Right** to reposition the Dock.

Ⓓ Click **Minimize Using** and then click **Genie Effect** or **Scale Effect**.

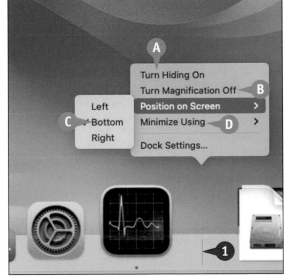

TIP

How else can I customize the Dock?
You can increase or decrease the size of the Dock by clicking the Dock divider bar and dragging it up or down. For more precise control of the Dock, press Control+click the Dock divider bar and then click **Dock Settings** to display the Desktop & Dock pane in System Settings. Here you can change the Dock size, turn on and adjust magnification, set the Dock's position, and choose the effect for minimizing windows. You can also choose other options for controlling the Dock's appearance and behavior.

Configure Control Center and the Menu Bar

You can configure the items that appear on the right side of the macOS menu bar between the last menu of the active app and the date and time readouts. Typically, you would want to put on the menu bar only those items to which you need instant access.

Some of the available items — such as Wi-Fi, Bluetooth, AirDrop, and Focus — always appear in Control Center, giving you easy access to them at all times. You can decide whether other items appear in Control Center.

Configure Control Center and the Menu Bar

1 Click **System Settings** (⚙) on the Dock.

The System Settings window opens.

2 Click **Control Center** (▣) in the sidebar.

The Control Center pane appears.

Ⓐ The Control Center Modules section shows modules that always appear in Control Center. You can choose which of them to display in the menu bar as well.

3 Click the pop-up menu (⬍) for each feature, and then click **Show in Menu Bar** or **Don't Show in Menu Bar**, as needed.

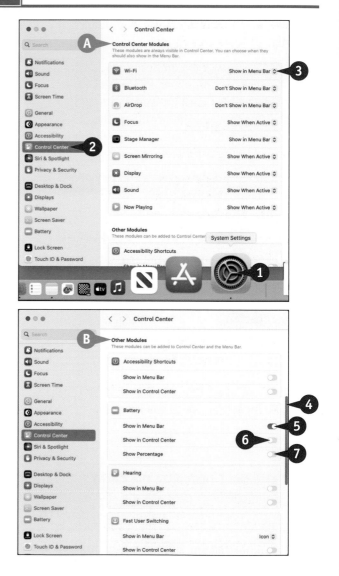

4 Scroll down.

Ⓑ The Other Modules section shows modules that you can add to Control Center, to the menu bar, or to both.

5 For each module, set the **Show in Menu Bar** switch to On (◯) to display the module in the menu bar.

6 Set the **Show in Control Center** switch to On (◯) to display the module in Control Center.

7 For the Battery module, set the **Show Percentage** switch to On (◯) to display the remaining battery percentage on the battery icon.

8 Scroll down.

C The Menu Bar Only section contains modules that appear only in the menu bar.

9 Click the **Spotlight** pop-up menu (◊), the **Siri** pop-up menu (◊), and the **Time Machine** pop-up menu (◊), and then click **Show in Menu Bar** or **Don't Show in Menu Bar**, as needed.

10 Click the **Automatically hide and show the menu bar** pop-up menu (◊), and then click **Always**, **On Desktop Only**, **In Full Screen Only**, or **Never**, as needed.

11 Click the **Recent documents, applications, and servers** pop-up menu (◊), and then click **None**, **5**, **10**, **15**, **20**, **30**, or **50**.

12 Click **Clock Options**.

The Clock Options dialog opens.

13 Click the **Show date** pop-up menu (◊), and then click **When Space Allows**, **Always**, or **Never**, as needed.

14 Set the **Show the day of the week** switch to On (⬤) or Off (⬤), as needed.

15 On the Style row, click **Digital** (◯ changes to ⬤) or **Analog** (◯ changes to ⬤), as needed.

16 Set the **Show AM/PM** switch to On (⬤) or Off (⬤), as needed.

17 Click **Done** to close the Clock Options dialog.

18 Click **Close** to close System Settings.

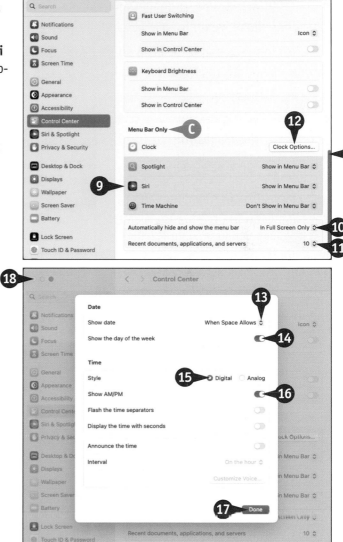

TIP

What are the Accessibility Shortcuts?
The Accessibility Shortcuts are quick-access tools for a wide range of Accessibility features. The Vision category contains the VoiceOver, Zoom, Invert Colors, Color Filters, Increase Contrast, and Reduce Transparency features. The Physical and Motor category contains the Sticky Keys, Slow Keys, Mouse Keys, Full Keyboard Access, Accessibility Keyboard, and Head Pointer features.

Add or Remove Desktop Spaces

macOS enables you to create multiple desktop spaces on which to arrange your documents and apps. You can switch from space to space quickly to move from app to app. You can tie an app to a particular space so that it always appears in that space or allow it to appear in any space.

When you no longer need a desktop space, you can remove it in just moments. To configure desktop spaces, you use Mission Control.

Add or Remove Desktop Spaces

1 Swipe up the trackpad with three fingers or press `Control` + `↑` (not shown).

Note: On some keyboards, you press `F9` to invoke Mission Control. On other keyboards, you press `F3`. Depending on your MacBook's settings, you may need to press `F3` in combination with the function key.

Note: Another way to invoke Mission Control is to run the Mission Control app from Launchpad or from an icon you add to the Dock.

The Mission Control screen appears.

Ⓐ You can click a window to switch quickly to it.

2 Move the pointer to the top of the screen.

The bar at the top of the screen grows deeper when the pointer is over it.

A panel showing a + sign appears.

Note: If you have positioned the Dock on the right, the + sign appears in the upper-left corner of the screen.

3 Click the + panel.

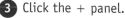

B Another desktop space appears at the top of the Mission Control screen.

C You can click a window and drag it to the desktop space in which you want it to appear.

4 Click a window you want to display full screen, and then drag it to the bar at the top of the screen.

The app appears as a full-screen item on the row of desktops.

D When you need to close a desktop, move the pointer over it, and then click **Close** (ⓧ).

5 Click the desktop space or full-screen app you want to display.

The desktop space or app appears.

TIP

How can I assign an app to a particular desktop?
First, use Mission Control to activate the desktop to which you want to assign the app. Then press Control+ click or right-click the app's Dock icon, click **Options**, and click **This Desktop**. If you want to use the app on all desktops, click **All Desktops** in the Assign To section of the Options continuation menu.

Create Hot Corners to Control Screen Display

The Hot Corners feature enables you to trigger actions by moving the pointer to the corners of the screen. You can set up hot corners for as many of the four corners of the screen as you want. Each hot corner can perform an action such as opening Mission Control, displaying your desktop, or starting the screen saver.

To set up hot corners, you use the Hot Corners dialog. You can open this dialog from the Desktop & Dock pane in System Settings.

Create Hot Corners to Control Screen Display

Set Up a Hot Corner

1 Click **System Settings** (⚙) on the Dock.

The System Settings window opens.

2 Click **Desktop & Dock** (▣).

The Desktop & Dock pane appears.

3 Scroll down to the Mission Control section at the bottom of the pane.

4 Click **Hot Corners**.

The Hot Corners dialog opens.

5 Click the pop-up menu (⬍) for the corner you want to configure.

The pop-up menu opens.

6 Click the action you want to assign to the hot corner.

7 Choose other hot corner actions as needed.

Note: You can set up multiple hot corners for the same feature if you want.

8 Click **Done**.

The Hot Corners dialog closes.

9 Click **System Settings** on the menu bar.

The System Settings menu opens.

10 Click **Quit System Settings**.

System Settings closes.

Using a Hot Corner to Run Mission Control

1 Move the pointer to the hot corner you allocated to Mission Control.

The Mission Control screen appears.

2 Click the window you want to display.

The window becomes active.

TIP

Are there other ways I can run Mission Control?
Yes. You can also run Mission Control by using a trackpad gesture or a keystroke or key combination. Click **System Settings** (⚙) on the Dock to open System Settings. To configure a gesture, click **Trackpad** (▣) in the sidebar, click the **More Gestures** tab, click the **Mission Control** pop-up menu (◇), and then click the gesture, such as **Swipe Up with Three Fingers**. To set up a keystroke or key combination to run Mission Control, click **Desktop & Dock** (▣) in the sidebar, click **Shortcuts** at the bottom of the pane, and then work in the Shortcuts dialog that opens. Press and hold ⌘, **Option**, **Control**, **Shift**, or a combination of the four keys to add them to the keystroke.

Make the Screen Easier to See

The Accessibility features in macOS include several options for making the contents of your MacBook's screen easier to see. You can invert the colors, use grayscale instead of colors, enhance the contrast, and increase the cursor size. You can also turn on the Zoom feature to enable yourself to zoom in quickly up to the limit you set. To configure these options, you open System Settings and work in the Accessibility pane.

Make the Screen Easier to See

1 Click **System Settings** (⬤) on the Dock to open the System Settings window.

2 Click **Accessibility** (⬤) in the sidebar to display the Accessibility pane.

3 Click **Zoom** (⬤).

The Zoom pane appears.

4 Set the **Use keyboard shortcuts to zoom** switch to On (⬤).

5 Set the **Use trackpad gesture to zoom** switch to On (⬤).

6 Set the **Use scroll gesture with modifier keys to zoom** switch to On (⬤) to zoom by holding a modifier key and scrolling with the trackpad.

7 Click the **Modifier key for scroll gesture** pop-up menu (⌄), and then click **Control**, **Option**, or **Command**.

8 Click the **Zoom style** pop-up menu (⬍), and then click **Full screen**, **Split screen**, or **Picture-in-picture**, as needed.

9 If your MacBook uses multiple displays, click **Choose Display** and then click the display to use for zoom.

10 Click **Advanced** to open the Advanced dialog.

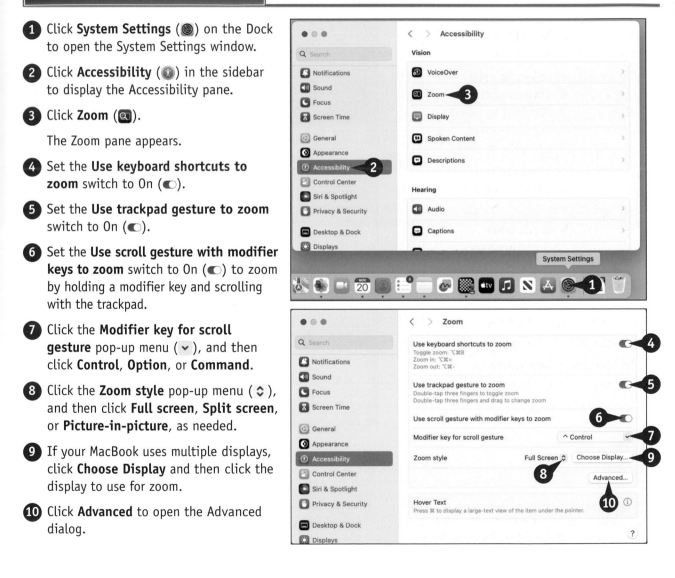

⑪ In the Zoomed Image Moves area, click **Continuously with Pointer** (◯ changes to ◉), **When Pointer Reaches Edge** (◯ changes to ◉), or **To Keep Pointer Centered** (◯ changes to ◉), to specify what happens when you move the pointer with zoom enabled.

⑫ Set the **Restore zoom factor on startup** switch to On (⬤) to restore the zoom when you log in.

⑬ Set the **Invert colors in split screen and picture-in-picture style** switch to On (⬤) to invert colors when using split-screen zoom or picture-in-picture zoom.

⑭ Set the **Smooth images** switch to On (⬤) to have macOS smooth zoomed images.

⑮ Set the **Flash screen when notification banner appears outside zoom view** switch to On (⬤) to get alerts for banners that appear outside the zoom view.

⑯ Click **OK** to close the Advanced dialog.

⑰ Set the **Hover Text** switch to On (⬤) to enable hover text.

⑱ Click **Options** (ⓘ).

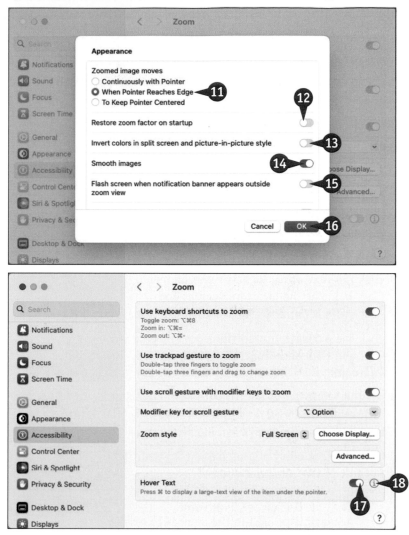

TIP

What is the quickest way to turn on the Universal Access features for seeing the screen?
Use keyboard shortcuts. Press `Option`+`⌘`+`8` to toggle zoom on or off. With zoom turned on, press `Option`+`⌘`+`=` to zoom in by increments. Press `Option`+`⌘`+`-` to zoom out by increments. If you enable the Smooth Images feature by selecting **Smooth images** (☑) in the Zoom options, you can press `Option`+`⌘`+`\` to toggle Smooth Images on or off.

continued ▶

The Hover Text feature enables you to quickly zoom an item by moving the mouse pointer over it and pressing ⌘ or another key you specify. You can configure options for text entry using Hover Text.

If you tend to lose track of the pointer, you can turn on the Shake Mouse Pointer to Locate feature, which expands the pointer when you move the pointer back and forth quickly. You can also increase the cursor size to make the pointer more visible.

Make the Screen Easier to See (continued)

The Options dialog for Hover Text opens.

19 Click the **Text size** spin buttons (⇕) to set the text size.

20 Click the **Text font** pop-up menu (⇕) and specify the font.

21 Click the **Text-entry location** pop-up menu (⇕) and click the location, such as **Top Left** or **Near Current Line**.

22 Click the **Activation modifier** pop-up menu (⇕) and then click **Control**, **Option**, or **Command**.

23 Set the **Triple-press modifier to set activation lock** switch to On (⬤) to lock Hover Text by triple-pressing the activation modifier key.

24 Click the **Text color** pop-up menu (⇕) and click **Default** or the color you want.

25 Click **Done** to close the Options dialog.

26 Click **Accessibility** (ⓘ) in the sidebar to display the Accessibility pane.

27 Click **Display** (🖥) to switch to the Display pane (not shown).

28 Set the **Invert colors** switch to On (⬤) if you want to invert the video colors for visibility.

29 On the Invert Colors Mode row, click **Smart** (○ changes to ◉) for smart color inversion, or **Classic** (○ changes to ◉) for standard color inversion.

Note: Smart inversion does not invert the colors of images.

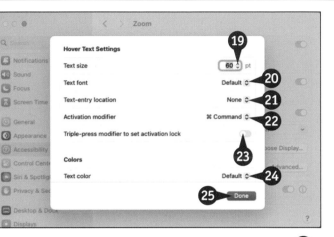

30 Set the switches for other options to On (⬤) or Off (○), as needed. See the tip for details.

31 Drag the **Display contrast** slider to set the display contrast.

32 Scroll down to the Text section.

33 Click **Text size** to display the Text Size dialog, drag the slider to set the default text size, choose which apps should use that size, and then click **Done**.

34 On the Menu Bar Size row, click **Default** (○ changes to ◉) or **Large** (○ changes to ◉), as needed.

35 Set the **Prefer horizontal text** switch to On (⬤) if you want to encourage languages that support vertical text to lay text out horizontally.

36 Set the **Shake mouse pointer to locate** switch to On (⬤) to be able to enlarge the pointer temporarily by shaking it.

37 Drag the **Pointer size** slider to set the pointer size.

38 Click **Pointer outline color** and click the outline color.

39 Click **Pointer fill color** and click the fill color.

Ⓐ You can click **Reset Colors** to reset the pointer colors.

40 Scroll down to the Color Filters section.

41 Set the **Color filters** switch to On (⬤) if you want to apply a filter.

42 Click the **Filter type** pop-up menu (⇕) and click the filter, such as **Color Tint**.

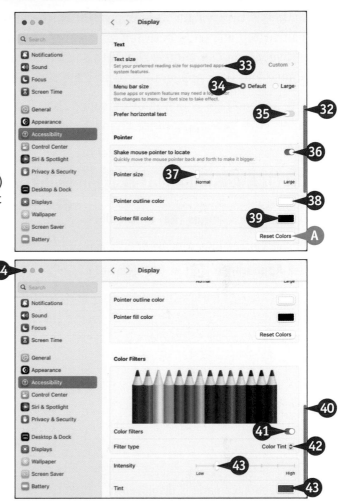

43 If controls for that filter appear, adjust them as needed.

44 Click **Close** (⬤) to close System Settings.

TIP

What do Reduce Motion, Dim Flashing Lights, Increase Contrast, Reduce Transparency, and Differentiate Without Color do?

Enable **Reduce motion** (⬤) to reduce 3D effects in the macOS user interface. Enable **Dim flashing lights** (⬤) to have macOS dim repeated flashing or strobing lights in videos. Enable **Increase contrast** (⬤) to increase the degree of contrast in the interface. Enable **Reduce transparency** (⬤) to reduce transparency so that items in the foreground show less of the items behind them. Enable **Differentiate without color** (⬤) to use noncolor means of differentiating interface items that are differentiated using color by default.

Configure the Appearance, Accent, and Highlight

macOS has two system-wide appearances, the Light appearance and the Dark appearance. You can switch manually between appearances by using the Appearance setting in System Settings. Alternatively, you can choose the Auto setting to have macOS switch automatically between Light appearance and Dark appearance to suit the time of day at your current location.

The Dark appearance, which is often called Dark Mode, can be easier on your eyes. If you use Dynamic wallpapers, the Dark appearance changes the wallpaper between the light still image and the dark still image.

Configure the Appearance, Accent, and Highlight

① Click **System Settings** (⚙) on the Dock.

The System Settings window opens.

② Click **Appearance** (◐) in the sidebar.

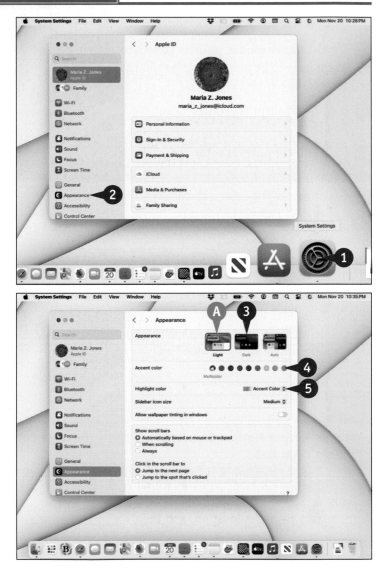
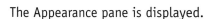

The Appearance pane is displayed.

Ⓐ In the Appearance area, the current appearance has a blue outline.

③ Click **Light**, **Dark**, or **Auto** to set the appearance.

④ In the Accent Color area, click the accent color you want to use.

⑤ Click the **Highlight color** pop-up menu (↕), and then click **Accent Color** or a preset color, such as **Blue**. Alternatively, click **Custom** and use the color picker to specify your preferred color.

6 Click the **Sidebar icon size** pop-up menu (⬍) and click **Small**, **Medium**, or **Large**, as you prefer.

7 Set the **Allow wallpaper tinting in windows** switch to On (⬤○) if you want the wallpaper color to show through light colors in windows.

8 Click **Close** (⬤).

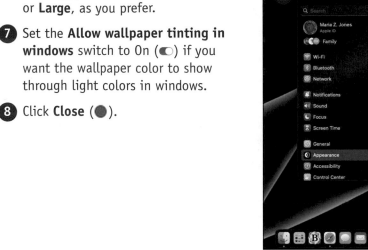

System Settings closes.

Ⓑ If you have a Dynamic wallpaper set, the wallpaper changes to suit the appearance.

TIP

How do I configure the time at which Auto changes the appearance?
Unlike on iOS devices, you cannot directly configure the schedule for changing the appearance; macOS uses the system time, the location, and the time of the year to set the schedule.

Configure the Keyboard

You can customize your MacBook's built-in keyboard or an external keyboard you connect to your MacBook via Bluetooth or USB. Your customization choices include adjusting the repeat rate and the delay until repeating starts and configuring the actions assigned to modifier keys such as ⌘, Option, and 🌐. You can also set up keyboard shortcuts for various macOS features and apps, and you can create text shortcuts that work for all devices using your Apple ID.

Configure the Keyboard

1 Click **System Settings** (⚙) on the Dock.

The System Settings window opens.

2 Click **Keyboard** (⌨) in the sidebar.

The Keyboard pane appears.

3 Drag the **Key repeat rate** slider to control how quickly a key repeats.

4 Drag the **Delay until repeat** slider to set the repeat delay.

5 If it appears, set the **Adjust keyboard brightness in low light** switch to On (⬤) to have your MacBook adjust the keyboard brightness automatically.

6 Drag the **Keyboard brightness** slider to set the keyboard brightness manually.

7 To turn the keyboard backlight off automatically, click the **Turn keyboard backlight off after inactivity** pop-up menu (⬍) and click the delay, such as **After 10 Seconds**.

8 Click the **Press** 🌐 key to pop-up menu (⬍), and then click **Change Input Source**, **Show Emoji & Symbols**, **Start Dictation (Press** 🌐 **Twice)**, or **Do Nothing**, as needed.

9 Set the **Keyboard navigation** switch to On (⬤) if you want to press Tab to move the focus in windows.

10 Click **Keyboard Shortcuts**.

The Keyboard Shortcuts dialog opens.

11 In the sidebar, click the category of shortcuts you want to configure.

This example uses **Modifier Keys**. This category enables you to configure Caps lock, Control, Option, ⌘, and 🌐 for either your MacBook's built-in keyboard or an external keyboard.

12 Click the **Select Keyboard** pop-up menu (⇕), and then select either **Apple Internal Keyboard/Trackpad** or the external keyboard.

13 On each row, click the pop-up menu, and then click the key action or **No Action**.

14 Click **Done**.

15 Click **Text Replacements**.

The Text Replacements dialog opens.

16 To add a text replacement, click **Add** (✚), type the term and its replacement in the dialog that opens, and then click **Add**.

17 To remove a text replacement, click it, and then click **Remove** (━).

18 Click **Done**.

The Text Replacements dialog closes.

19 Click **Close** (●).

System Settings closes.

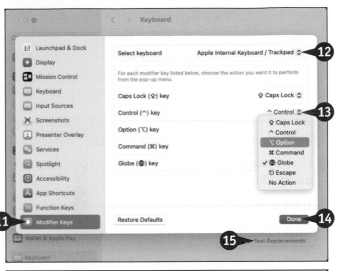

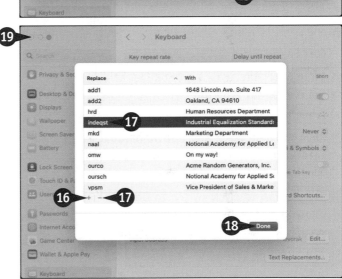

TIP

How do I add a different layout for my MacBook's keyboard?
Click **System Settings** (⚙) to open the System Settings window, click **Keyboard** (▭) in the sidebar to display the Keyboard pane, and then click **Edit** in the Text Input area. In the dialog that opens, click **Add** (✚), select the language and layout, and then click **Add**. Back in the first dialog, you can click your previous input method and click **Remove** (━) to remove it. If you prefer to keep both methods, set the **Show Input menu in menu bar** switch to On (🔵), click **Done** to close the dialog, and then click **Close** (●) to close System Settings. You can then use the input menu on the menu bar to switch keyboard layouts.

Configure the Trackpad or Other Pointing Device

You can customize the settings for your MacBook's trackpad to make it work the way you prefer. For example, you can adjust the firmness of the click, increase or decrease the tracking speed, and select which gestures to use. If you find the trackpad awkward, you can also connect a mouse or other pointing device via Bluetooth or USB; alternatively, you can turn on the Mouse Keys feature, which enables you to control the pointer from the keyboard, or another form of pointer control.

Configure the Trackpad or Other Pointing Device

① Click **System Settings** (⚙) on the Dock to open the System Settings window.

② Click **Trackpad** (🔲) in the sidebar to display the Trackpad pane.

③ Click **Point & Click** to display the Point & Click tab.

④ Drag the **Tracking speed** slider to adjust the tracking speed.

⑤ Drag the **Click** slider to set Light, Medium, or Firm clicks.

⑥ Set the **Silent clicking** switch to On (🔵) to turn off click noises.

⑦ Set the **Force Click and haptic feedback** switch to On (🔵) to enable the Force Click feature and vibration feedback.

⑧ Set the **Tap to click** switch to On (🔵) to enable clicking by tapping the trackpad.

⑨ Click the **Look up & data detectors** pop-up menu (⇕) and click the action you want.

⑩ Click the **Secondary click** pop-up menu (⇕) and click the action you want.

⑪ Click **Scroll & Zoom** to display the Scroll & Zoom tab.

⑫ Set the **Natural Scrolling** switch to On (🔵) to have scrolling follow your finger movements.

⑬ Set the **Zoom in or out** switch to On (🔵) to zoom by pinching in or out.

⑭ Set the **Smart zoom** switch to On (🔵) to zoom by double-tapping with two fingers.

⑮ Set the **Rotate** switch to On (🔵) to rotate objects by placing two fingers and rotating them.

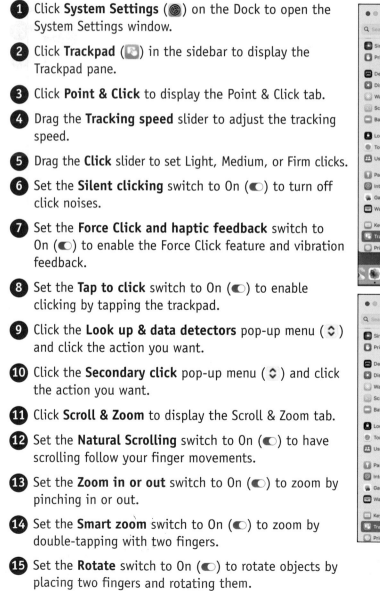

16 Click **More Gestures** to display the More Gestures tab.

17 Click the **Swipe between pages** pop-up menu ($\updownarrow$) and click the gesture.

18 Click the **Swipe between full-screen applications** pop-up menu ($\updownarrow$) and click the gesture.

19 Set the **Notification Center** switch to On (⬤) to open Notification Center by swiping left with two fingers from the right edge of the screen.

20 Click the **Mission Control** pop-up menu ($\updownarrow$) and click the gesture.

21 Click the **App Exposé** pop-up menu ($\updownarrow$) and click the gesture.

22 Set the **Launchpad** switch to On (⬤) to open Launchpad by pinching with your thumb and three fingers.

23 Set the **Show Desktop** switch to On (⬤) to display the desktop by spreading out with your thumb and three fingers.

24 Click **Accessibility** (⬤) in the sidebar to display the Accessibility pane.

25 Click **Pointer Control** (🔺) to display the Pointer Control pane (not shown).

26 Drag the **Double-click speed** slider to set the double-click speed.

27 Set the **Spring-loading** switch to On (⬤) to use spring-loading. See the tip.

A You can click **Trackpad Options** to configure further settings.

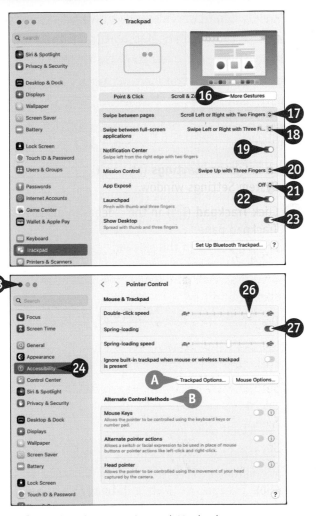

B In the Alternate Control Methods area, you can enable and configure Mouse Keys, Alternate Pointer Actions, or a Head Pointer.

28 Click **Close** (⬤) to close System Settings.

TIP

In the Pointer Control pane in Accessibility settings, what is spring-loading?

In Finder, when you drag files to a closed folder, the folder stays closed for a moment to enable you to drop the files without opening the folder. If you do not drop the files, the folder springs open after a short delay, revealing its contents so that you can navigate further. This action is called *spring-loading*; you can enable or disable it by setting the **Spring-loading** switch to On (⬤) or Off (⬜). With spring-loading enabled, you can reduce or increase the delay by dragging the **Spring-loading speed** slider.

Configure iCloud Settings

Apple's iCloud service adds powerful online sync features to your MacBook. iCloud also enables you to use the Find My Mac feature or the Find My app on iPhone or iPad to locate your MacBook if it goes missing — or even to erase the MacBook remotely.

To use iCloud, you set your MacBook user account to use your Apple ID and then choose which features to use. If you added your iCloud account when first setting up your MacBook, iCloud is already configured, but you may want to select different settings for it.

Configure iCloud Settings

1 Click **System Settings** (⚙) on the Dock.

The System Settings window opens.

2 Click **Apple ID** in the sidebar.

The Apple ID pane appears.

3 Click **iCloud** (☁).

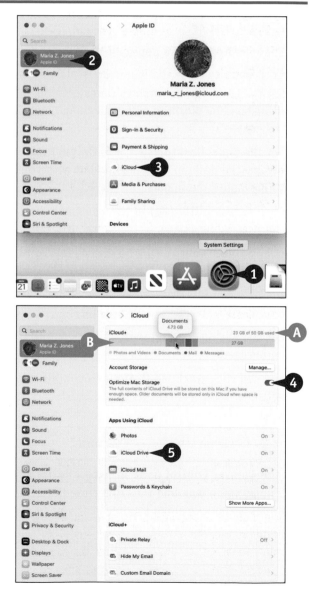

The iCloud pane appears.

Ⓐ The iCloud readout or iCloud+ readout shows how much storage space you have and how much you have used.

Note: iCloud is the free tier of the service. iCloud+ is the paid tier.

Ⓑ The histogram shows a visual representation of what is taking up space. Move the pointer over a color to display a pop-up label showing what the color represents. For example, yellow represents Photos and Videos, orange represents Documents, royal blue represents Mail, and green represents Messages.

4 Set the **Optimize Mac Storage** switch to On (⬤) to store as much of your iCloud Drive on your MacBook as possible.

5 Click **iCloud Drive** (☁).

The iCloud Drive dialog opens.

6 Set the **Sync This Mac** switch to On (⬤○) to sync iCloud files with your MacBook.

7 Set the **Desktop & Documents Folders** switch to On (⬤○) to store your Desktop folder and Documents folder in iCloud.

C You can click **Apps syncing to iCloud Drive** to open a dialog that shows all the apps currently syncing data to iCloud. You can set an app's switch to Off (○) to disable syncing. Click **Back** to return to the iCloud Drive dialog.

8 Click **Done** to close the iCloud Drive dialog.

9 Click **Manage**.

The Account Storage dialog opens.

D You can click **Change Storage Plan** to display the Upgrade iCloud dialog, which enables you to upgrade your storage to a paid level of iCloud+ or to downgrade it again.

E You can click a button to see details of an app's usage of iCloud and any actions you can take to change the usage.

10 Click **Done** to close the Account Storage dialog.

11 Click **Close** (⬤) to close System Settings.

TIP

How do I change my iCloud password?

Click **System Settings** (⚙) to open System Settings, click **Apple ID** to display the Apple ID pane, and then click **Sign-In & Security** (🔒) to display the Sign-In & Security pane. Click **Change Password**, and then follow the prompts.

In the Sign-In & Security pane, you can verify that Two-Factor Authentication is enabled for your iCloud account, or enable it if it is not yet enabled. You can also set up the Sign In with Apple feature, configure Account Recovery to enable yourself to recover from getting shut out of your account, or specify a Legacy Contact who can access your account after your death.

Add a Second Display

macOS enables you to add one or more external displays to your MacBook to give yourself more space for your apps. You connect the display to one of the USB-C/Thunderbolt ports on the MacBook. If the display has a USB-C connector, you connect it directly; if not, you use a suitable adapter. Some displays that connect via USB-C can charge your MacBook; some provide additional ports.

You use the Displays pane in System Settings to specify the external display's role, such as an extended display; to set the resolution; and to specify the arrangement of the displays.

Add a Second Display

1 Connect the display to your MacBook (not shown).

Note: To connect an external display to the MacBook, you may need Apple's USB-C Digital AV Multiport Adapter or a functional equivalent. This adapter provides an HDMI port for connecting the external display.

2 Connect the display to power and turn it on (not shown).

3 Click **System Settings** () on the Dock to open the System Settings window.

4 Click **Displays** (🖥️) in the sidebar to show the Displays pane.

5 Click the display you added.

The controls for the display appear.

6 Click the **Use as** pop-up menu (‹›) and click **Main display**, **Extended display**, or **mirror for Built-in Display**, as needed.

7 Click the resolution you want to use.

A If the resolution you want does not appear, set the **Show all resolutions** switch to On (◉).

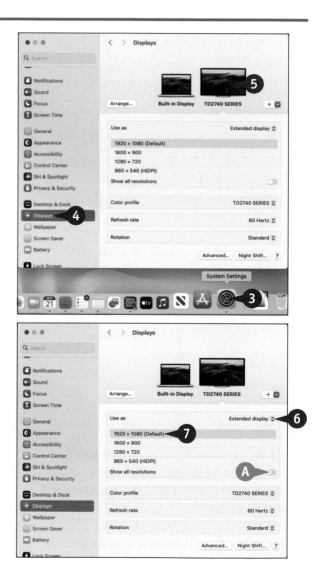

8 If you need to apply a different color profile to the display, click **Color Profile** (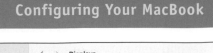), and then click the appropriate profile.

9 If you need to adjust the display's refresh rate, click **Refresh Rate** (⬦), and then click the appropriate rate.

10 If you need to change the rotation, click **Rotation** (⬦), and then click **Standard**, **90°**, **180°**, or **270°**, as needed.

11 Click **Arrange**.

The Arrange Displays dialog opens.

B To identify a display, hold the pointer over it.

12 Drag the display thumbnails to tell macOS the displays' positions relative to each other.

13 To move the menu bar and Dock, drag the menu bar from the current thumbnail to the other thumbnail.

14 Click **Done**.

The Arrange Displays dialog closes.

15 Click **Close** (●).

System Settings closes.

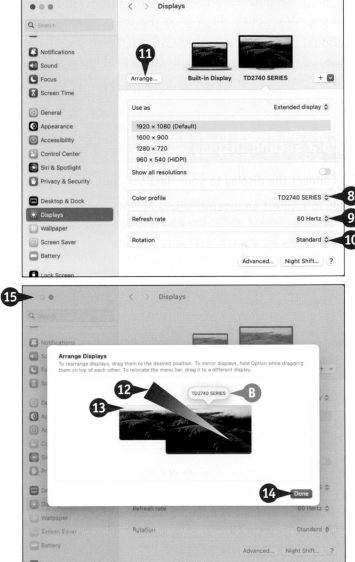

TIP

Can I add two external displays?
It depends on the MacBook model. The MacBook Air models and the M3 MacBook Pro support only one external display, but it can be of up to 6K resolution. The M3 Pro MacBook Pro supports one external display of up to 8K or two external displays of up to 6K. The M3 Max MacBook Pro supports up to four external displays — three at 6K over Thunderbolt plus one at 4K over HDMI. Alternatively, the M3 Max supports two 6K displays via Thunderbolt and one 8K display via HDMI. Check the specifications for your MacBook model to verify how many external displays it can drive.

Using an Apple TV and HDTV as an Extra Display

With a second-generation or later Apple TV — any black model — you can wirelessly broadcast your MacBook's display on the device to which the Apple TV is connected. This is great for watching movies or videos on a big-screen TV, enjoying a shared web-browsing session, or giving presentations from your MacBook to a group of people. To broadcast to an Apple TV, your MacBook uses AirPlay. This technology enables Macs, iPhones, and iPads to send a signal to an Apple TV for it to display on a television.

Using an Apple TV and HDTV as an Extra Display

Note: If the Apple TV is not already set up, connect it to a power outlet, to your television, and to your wireless network. Enable AirPlay by opening the Apple TV's Settings screen, selecting **AirPlay**, and setting AirPlay to On.

1 Click **Wi-Fi** (🛜) on the menu bar.

The Wi-Fi menu opens.

2 Click the network to which Apple TV is connected.

Note: Your MacBook and the Apple TV must be on the same network for AirPlay to work.

Ⓐ Your MacBook connects to the network.

3 Click **Screen Mirroring** (🖵) on the menu bar.

The Screen Mirroring menu opens.

4 Click the Apple TV you want to use.

Note: If the Enter the AirPlay Code dialog opens, type the AirPlay code displayed on the Apple TV. The dialog closes automatically after you have correctly entered the full code.

Your MacBook's desktop appears on the television to which the Apple TV is connected. By default, the Apple TV mirrors the MacBook's screen, showing the same image.

5 Click **Screen Mirroring** (⬚) on the menu bar.

The Screen Mirroring menu opens.

6 Click **Apple TV** if needed to expand the menu section for the Apple TV your MacBook is using.

7 Click **Use As Separate Display** if you want to use the HDTV as a separate display. Otherwise, leave the Mirror Built-In Retina setting selected.

macOS extends your desktop to the HDTV, giving you more desktop space.

8 When you finish using the HDTV, click **Screen Mirroring** (⬚) on the menu bar.

The Screen Mirroring menu opens.

9 Click the Apple TV you have been using.

Your MacBook stops displaying content via the Apple TV.

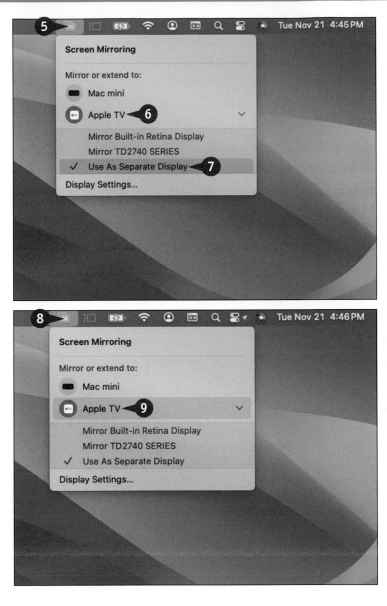

TIP

Why do I not see any AirPlay devices available on my MacBook?
First, check the network configuration of each device to make sure all devices are on the same network. Second, the network you are using may not support the protocols AirPlay uses; in some public areas, networks are designed to prevent streaming of content. If you cannot use a different network, see if the available one can be reconfigured to support AirPlay.

Using an iPad as an Extra Display and Input Device

The Sidecar feature enables you to use an iPad as an extra display for your MacBook Pro or MacBook Air, giving you extra space for your work. You can use Sidecar either by connecting both the MacBook and the iPad to the same Wi-Fi network or by connecting them with a USB cable. With both devices logged in to the same iCloud account, you can quickly move windows between the MacBook's screen and the iPad's screen, as needed. You can also show the sidebar on the iPad. If your MacBook has the Touch Bar, you can display that on the iPad, too.

Using an iPad as an Extra Display and Input Device

1 Click **System Settings** (⚙) on the Dock to open the System Settings window.

2 Click **Displays** (▦) in the sidebar to show the Displays pane.

3 Click **Add** (◆) to open the pop-up menu.

4 Click the iPad.

The Displays pane shows the internal display and the iPad.

5 Click the iPad to display its controls.

6 Click the **Use as** pop-up menu (↕) and click **Main display**, **Extended display**, or **Mirror for Built-in Display**, as needed.

7 Click the **Show Sidebar** pop-up menu (↕) and click **Never**, **On the left**, or **On the right** to specify whether and where to show the sidebar.

8 Click the **Show Touch Bar** pop-up menu (↕) and click **Never**, **On the top**, or **On the bottom** to specify whether and where to show the Touch Bar if your MacBook includes it.

9 Set the **Enable double tap on Apple Pencil** switch to On (◉) or Off (○), as needed. See the second tip.

10 Click **Arrange**.

The Arrange Displays dialog opens.

Ⓐ To identify a display, hold the pointer over it.

⑪ Drag the display thumbnails to tell macOS the displays' positions relative to each other.

Ⓑ You can move the menu bar and Dock by dragging the menu bar from the current thumbnail to the other thumbnail.

⑫ Click **Done**.

The Arrange Displays dialog closes.

⑬ Click **Advanced** to open the Advanced dialog.

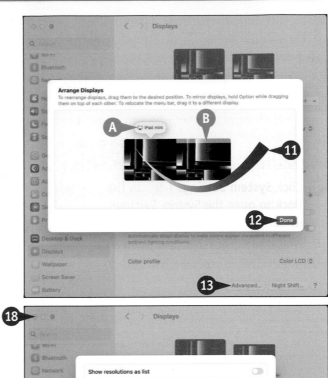

⑭ Set the **Allow your pointer and keyboard to move between any nearby Mac or iPad** switch to On (⬤).

⑮ Set the **Push through the edge of a display to connect a nearby Mac or iPad** switch to On (⬤) to establish a connection by moving the pointer to the edge of the display.

⑯ Set the **Automatically reconnect to any nearby Mac or iPad** switch to On (⬤) if you want to use automatic reconnection.

⑰ Click **Done** to close the Advanced dialog.

⑱ Click **Close** (⬤) to close System Settings.

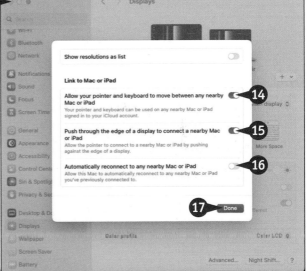

TIPS

How do I disconnect the iPad from my MacBook?

Tap **Disconnect** (▧) at the bottom of the sidebar on the iPad, and then tap **Disconnect** in the dialog that opens.

What does Enable Double Tap on Apple Pencil let me do?

This feature enables you to double-tap on the side of the Apple Pencil to perform custom actions in apps. Only some apps support custom actions, and the actions themselves vary from app to app. Either experiment by double-tapping the Apple Pencil or consult the documentation for the apps you use.

Connect External Devices

To extend your MacBook's capabilities, you can connect a wide variety of external devices. This section covers connecting speakers for audio output, connecting printers for creating hard-copy output from digital files, and connecting external drives for extra storage.

Connect Speakers

Each MacBook model has built-in speakers, but you can connect external speakers when you need greater volume. The easiest way to connect speakers is via the 3.5mm headphone jack that each MacBook model includes. On some models, the headphone jack can output only an analog signal. On other models, the headphone jack can output either an analog signal or a digital signal and switches automatically between analog and digital to match the type of cable you connect.

You can also connect speakers via a USB port. For any current MacBook, you need Apple's USB-C Digital AV Multiport Adapter or a similar adapter to provide a regular USB port instead of the USB-C port.

If you prefer not to connect your MacBook to speakers via a cable, you can use AirPlay or Bluetooth instead. For AirPlay, you need either speakers that are AirPlay-capable or speakers connected either to an AirPort Express wireless access point or to an Apple TV. For Bluetooth, you need either compatible Bluetooth speakers or a compatible Bluetooth audio receiver.

Whichever means of connection you use, you can use the Output pane in Sound settings to specify which audio device to use.

For easy control of audio inputs and outputs as well as the volume, click **Control Center** (🎛️) in the sidebar. In the Control Center Modules section, click the **Sound** pop-up menu (⬍) and click **Always Show in Menu Bar**.

You can then press Option +click **Sound** (A, 🔊) to display a menu for selecting sound output and input options quickly. For example, you can click a device in the Output list on the menu to direct the sound output to that device. You can also click **Sound Settings** (B) to display the Sound pane in System Settings.

Connect a Printer

To print from your MacBook, you need to connect a printer and configure a *driver*, the software for the printer. macOS includes many printer drivers, so you may be able to connect your printer and simply start printing. But if your printer is a new model, you may need to locate and install the driver for it.

You can connect a printer directly to your MacBook by using a USB cable. Unless the printer has a USB-C cable, you will need Apple's USB-C Digital AV Multiport Adapter or a similar adapter to provide a regular USB port.

After connecting the printer to the MacBook and to power and turning on the printer, press **Control**+click **System Settings** (⬢) on the Dock and then click **Printers & Scanners** on the contextual menu to display the Printers & Scanners pane in System Settings. See if the printer appears in the Printers list (C). If not, click **Add Printer, Scanner, or Fax** (D) and use the Add dialog to add the printer.

Connect an External Drive

To give yourself more disk space, you can connect an external drive to your MacBook via one of its USB-C/Thunderbolt ports. If the drive has a USB-C cable, you can connect it directly; if not, you will need Apple's USB-C Digital AV Multiport Adapter or a similar adapter to provide a regular USB port.

The drive can connect to your MacBook using either a USB standard or a Thunderbolt standard. USB-connected drives are ubiquitous and affordable, whereas Thunderbolt drives offer better performance than USB can provide, but they typically cost more. You can connect a USB drive to any computer, but relatively few non-Mac computers support Thunderbolt.

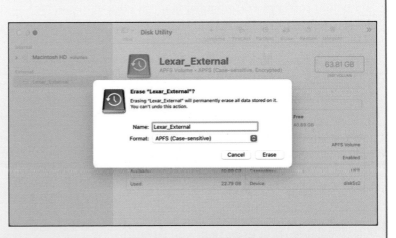

For best results, choose a drive that is designed for use with Macs. Normally, after connecting a drive, you will find that it appears in the Devices section of the sidebar in Finder windows; if so, you can simply start using the drive. But if you need to store large files on the drive, you may want to reformat the drive using the APFS file system. You can do this by using Disk Utility, which you can launch by clicking **Launchpad** (⊞) on the Dock, typing **disk**, and then clicking **Disk Utility** (🗄).

Explore Other Important Settings

macOS is highly configurable, and the System Settings app includes many settings beyond those you have met so far in this chapter. This section introduces you to four other categories of settings you may want to explore in order to get the most out of your MacBook: Language & Region, Extensions, Startup Disk, and Passwords.

To choose these settings, first display the System Settings window by either clicking **System Settings** (⚙) on the Dock or clicking **Apple** (🍎) on the menu bar and then clicking **System Settings** on the menu.

Choose Language & Region Settings

Click **General** (⚙) in the sidebar to display the General pane, and then click **Language & Region** (🌐) to display the Language & Region pane. Use the controls in the Preferred Languages pane to specify the languages you want to use, and then use the lower controls to specify the region, the calendar type, which temperature scale to use, the measurement system, the first day of the week, and date and number formats.

Click **Advanced** to open the Advanced dialog. Here, you can choose a wider variety of language and region settings. For example, you can click the **Dates** tab at the top and configure custom date formats as needed, or click the **Times** tab and set up exactly the time formats needed.

Choose Extensions Settings

Click **Privacy & Security** (🛡) in the sidebar, scroll down to the bottom of the Privacy & Security pane, and then click **Extensions** (🧩) to display the Extensions pane. Here, you can manage extensions, add-on components that provide additional functionality to macOS.

Click the category of extensions you want to configure. For example, click **Sharing** (A) to display the Select Extensions for Sharing with Others dialog, in which you can select (✓) or deselect (☐) the check boxes to specify which items you want to have on the Share menu.

Choose Startup Disk Settings

Click **General** (⚙️) in the sidebar to display the General pane, and then click **Startup Disk** (💾) to display the Startup Disk pane. Here, you can choose which disk to use for starting your MacBook. This functionality is useful not only for troubleshooting your MacBook and repairing its operating system, but also enabling you to switch among multiple operating systems installed on your MacBook.

Manage Your Stored Passwords

When you enter a password manually, you can have macOS store it so that it can enter the password for you automatically in the future. To manage your passwords, to update a password manually, or to look up a forgotten password, click **Passwords** (🔑) in the sidebar. In the Passwords pane, either click in the Search box (B) and type a partial name or keyword, or simply browse the list of passwords to find the one you want.

Click **Details** (ⓘ) to display the Details dialog for the password. You can click the password (C) to display its characters instead of the security-conscious dots. You can click **Change Password on Website** (D) to open a browser window to the website and start changing the password. You can click **Set Up** (E) to set up a verification code for the password. Or you can click **Edit** (F) to edit the user name, password, or notes manually. Click **Done** (G) when you finish.

CHAPTER 3

Sharing Your MacBook with Others

macOS makes it easy to share your MacBook with other people. Each user needs a separate user account for documents, e-mail, and settings.

Create a User Account

A user account is a group of settings that controls what a user can do in macOS. By creating a separate user account for each person who uses your MacBook regularly, you enable users to have their own folders for files and to use the settings they prefer. You can also use the Screen Time feature to limit and monitor the actions a child user can take.

When initially setting up your MacBook, you create an administrator account that you can use to configure macOS. You can also create a non-administrator account for yourself for day-to-day use.

Create a User Account

1. While logged in using your administrator account, press `Control`+click **System Settings** (⚙) on the Dock.

2. Click **Users & Groups**.

   The System Settings window opens.

   The Users & Groups pane appears.

3. Click **Add User**.

   The Users & Groups dialog opens, prompting you to enter your password.

4. Type your password.

5. Click **Unlock**.

   System Settings unlocks the settings.

   The New User dialog opens.

6. Click **New User** (⇕) and then click **Standard**.

7. In the Full Name box, type the user's name, such as **Kay** or **Kay Renner**.

8. If necessary, adjust the default account name that macOS derives from the user's name.

9. Click **Password Assistant** (🔑).

The Password Assistant dialog opens.

10 Click **Type** (‡) and click the type, such as **Letters & Numbers**.

11 Drag the **Length** slider to set the password length.

Ⓐ The Suggestion box shows the suggested password.

Ⓑ The Quality gauge shows the password's relative strength. Make sure the gauge shows green rather than yellow or red.

Ⓒ Password Assistant automatically enters the password in the Password box and the Verify box.

12 Optionally, type a password hint. See the second tip for details.

13 Click **Create User**.

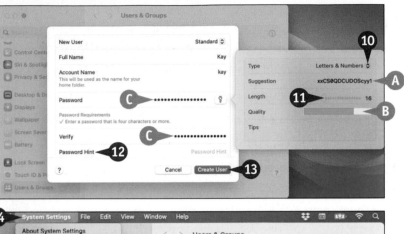

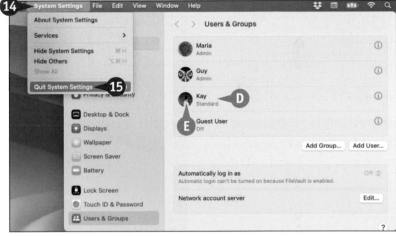

The New Account dialog closes.

Ⓓ The new account appears in the Users & Groups list.

Ⓔ You can click the default profile picture and then select a new profile picture or create a Memoji character, a 3-D avatar.

14 Click **System Settings**.

The System Settings menu opens.

15 Click **Quit System Settings**.

The System Settings app closes.

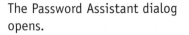

TIPS

Can I type a password instead of choosing one from Password Assistant?

Yes, you can type a password in the Password box. Consider creating a simple password and then having the user create a new password after they log in. This way, you do not know the user's password, making it more secure.

Should I create a password hint?

This is entirely your decision. Apple recommends creating a password hint, but it can be difficult to create a hint that will help the user remember the password but will not enable an attacker — who perhaps knows the user and follows their social media activities — to guess the password.

Configure Your MacBook for Multiple Users

As well as letting any user log in individually from the login screen, macOS includes a feature called *Fast User Switching* that enables multiple users to remain logged in to your MacBook at the same time. After you enable Fast User Switching, another user can log in either directly from your macOS session or from the login screen. Your macOS session keeps running in the background, ready for you to resume it.

You can also enable the Guest User account to give a visitor temporary use of your MacBook without creating a dedicated account.

Configure Your MacBook for Multiple Users

1 Click **Apple** (🍎).

The Apple menu opens.

2 Click **System Settings**.

The System Settings window opens.

3 In the sidebar, click **Control Center** ().

The Control Center pane appears.

4 Scroll down until the Other Modules heading appears at the top of the System Settings window.

Note: Fast User Switching uses more memory and resources, so it can make your MacBook run more slowly. If your MacBook runs too slowly, try turning Fast User Switching off by setting the **Show in Menu Bar** pop-up menu to Don't Show and the **Show in Control Center** switch to Off ( ).

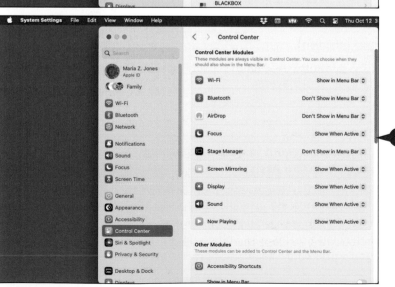

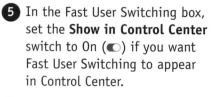

⑤ In the Fast User Switching box, set the **Show in Control Center** switch to On (⚪) if you want Fast User Switching to appear in Control Center.

⑥ Click the **Show in Menu Bar** pop-up menu (⇕).

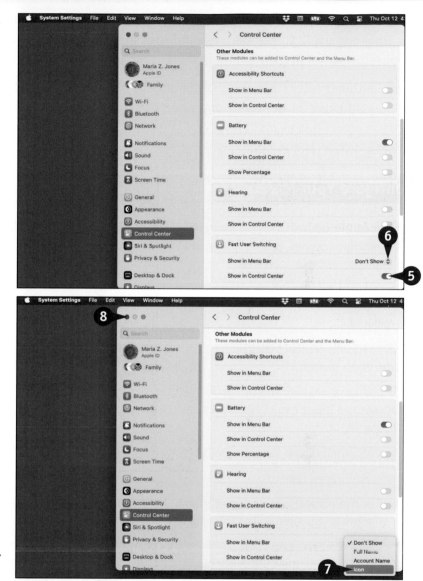

The pop-up menu opens.

⑦ Click **Full Name** to show usernames, click **Account Name** to show account names, or click **Icon** to show icons.

Note: Icon is the most compact of the three settings, so it is the best choice if your MacBook's menu bar is already full.

Fast User Switching is now enabled.

⑧ Click **Close** (⚫).

The System Settings window closes.

TIP

How do I set up the Guest User feature?

Click **Apple** (🍎) and **System Settings** to open the System Settings app, and then click **Users & Groups** (👥) to display the Users & Groups pane. On the Guest User row, click **Info** (ⓘ) to display the Info dialog for Guest User. Set the **Allow guests to log in to this computer** switch to On (⚪). Set the **Limit Adult Websites** switch to On (⚪) if you want to limit the guest user's access to adult websites. Set the **Allow guest users to connect to shared folders** switch to On (⚪) if you want to allow guests to access shared folders. Click **OK** to close the Info dialog.

Share Your MacBook with Fast User Switching

With Fast User Switching enabled, multiple users can remain logged in to macOS on your MacBook. Only one user can use the keyboard, trackpad, and screen at any given time, but each other user's computing session keeps running in the background, with all their applications still open.

macOS automatically stops multimedia playing when you switch users. For example, if another user is still playing music in the Music app when you switch to your user account, Music stops playing.

Share Your MacBook with Fast User Switching

Log in to the MacBook

Note: You can also log in using Touch ID by placing one of your registered fingertips on the fingerprint scanner. You do not need to click your username or icon first.

Note: To scroll the list left or right, swipe left or right with two fingers on the trackpad.

1 In the login window, click your username or icon.

A If you have enabled the Guest User account, a guest can click **Guest User** to log in.

macOS prompts you for your password.

B Your wallpaper appears.

Note: You can press Esc to return to the login window if you need to log in using a different account.

C If your MacBook has multiple input methods set up, choose the appropriate input method.

2 Type your password.

The password characters appear as dots to prevent anyone reading them.

3 Click **Log In** (⊝) or press Return.

macOS logs you in.

Your desktop appears.

Display the Login Window

1 When you are ready to stop using the MacBook but do not want to log out, click your name, account name, or icon on the menu bar.

The Fast User Switching menu opens.

D You — or another user — can click the user's name to log in that user.

2 Click **Login Window**.

The login window appears.

E Your username shows a check mark icon (☑), indicating that you have a session open.

Any of the MacBook's users can log in by clicking their username or by placing their registered Touch ID finger on the fingerprint scanner.

TIP

How can I log another user out so that I can shut down?

From the Fast User Switching menu, you can see what other users are logged in to the MacBook. If possible, ask each user to log in and then log out before you shut down. If you must shut down the MacBook and you are an administrator, click **Shut Down** (⏻) in the upper-right corner of the login window. macOS warns you that there are logged-in users. Type your name and password, and then click **Shut Down**.

Set Up Family Sharing

The Family Sharing feature lets you share purchases from the iTunes Store, the App Store, and the Apple Books service with your family members. You can also share an iCloud storage plan; subscriptions to Apple services, such as Apple One, Apple Music, Apple TV, Apple Arcade, and Apple News + ; and other items, such as a photo album and a family calendar. Family Sharing also enables family members to find each other's devices by using the Find My app.

Understanding How Family Sharing Works

The Family Sharing group consists of one organizer plus up to five family members. This section assumes that you are the organizer. The family members can be either adults or children.

Each family member must have an Apple ID, an e-mail address and password credential used to identify users to Apple services. Any family member aged 13 or older who does not yet have an Apple ID can create one online at https://appleid.apple.com or by using various mechanisms built into macOS, iOS, and iPadOS. You, the organizer, can create an Apple ID for a family member younger than 13.

You add the family members to Family Sharing by using their Apple IDs. You can either send an invitation via e-mail or, if the family member is present, have them enter their Apple ID details on a device you use for administering Family Sharing, such as your MacBook.

You can administer Family Sharing on a Mac, on an iPad, or on an iPhone.

For the services shared with family members, you can specify which members can purchase items, such as songs and movies, without restriction and which members need your approval. You can designate adult family members as parents/guardians who can approve requests to purchase content.

Access the Family Sharing Controls

You access the Family Sharing controls via the Family Sharing pane in the System Settings app. The quick way to open the Family Sharing pane is to `Control` +click **System Settings** (⚙) on the Dock and then click **Family** on the contextual menu. Alternatively, click **Apple** (🍎) and then **System Settings** to open the System Settings window, and then click **Family** (A). If the Family item does not appear in the System Settings window, click **Apple ID**, and then click **Family Sharing** (👥).

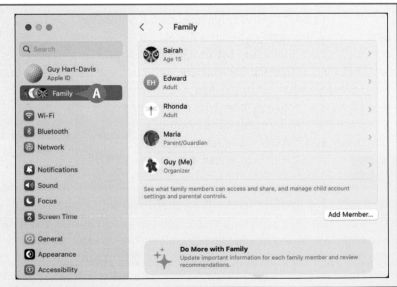

Add a Person to Family Sharing

To add a person to Family Sharing, click **Add Member** in the Family pane. In the Invite People to Your Family dialog, you can either add a person who has an Apple ID or set up a new Apple ID for a child who does not yet have one.

For a person with an Apple ID, click **Invite People** (B). In the Send Invitations dialog that opens, click **AirDrop** (C, 📡), **Mail** (D, ✉️), **Messages** (E, 💬), or **Invite in Person** (F, 👥), as needed. Click **Continue** (G), and then follow the prompts to identify the person and send the invitation.

For a child without an Apple ID, click **Create Child Account** (H) in the Invite People to Your Family dialog. Follow the prompts for authenticating yourself, providing the information needed to set up the child's account; and accepting terms and conditions, such as the Parental Consent Terms and Conditions.

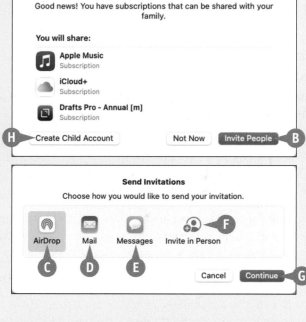

Complete the Family Checklist

In the lower part of the Family pane, click **Family Checklist** (☑) to open the Family Checklist dialog. This dialog presents a Settings for You list of iCloud Family features you have not yet configured. For example, you can click **Set Up a Recovery Contact** (I) in the Add a Recovery Contact box to designate a family member who can get you back into your account if you forget your password; or you can click **Add a Legacy Contact** in the Add a Legacy Contact box to specify a family member who can access your data after you die.

continued ▶

If you turn on the Ask to Buy feature for some of your family members, you as the organizer will receive notifications of their requests to buy items. You may want to designate some adult family members as parents/guardians who can approve or deny such requests.

After setting up Family Sharing, you can use the Screen Time feature to apply restrictions to what child family members can do on Macs, iPads, and iPhones on which they sign in using their Apple ID. The following sections explain how you can set "downtime" for a user, set limits on app and website usage, and apply content restrictions.

Configure Ask to Buy for a Family Member

To control whether a family member needs permission to purchase items and services from Apple, you configure the Ask to Buy setting. In the Family pane, click the family member to display their configuration dialog, and then click **Ask To Buy** (, A) in the sidebar to display the Ask to Buy controls. Set the **Require Purchase Approval** switch (B) to On (), and then click **Done** (C) to close the configuration dialog.

Designate an Adult Family Member as a Parent/Guardian

You can designate an adult family member as a parent/guardian who can approve or deny purchases from the services you enable in Family Sharing. In the Family pane, click the family member to display their configuration dialog, and then click **Parent/Guardian** (, D) in the sidebar to display the Parent/Guardian controls. Move the **Set as Parent/Guardian** switch (E) to On (), and then click **Done** (F) to close the configuration dialog.

To see the parents and guardians for a family member, click that family member. In the configuration dialog that opens, click **Parents/ Guardians** () in the sidebar to display the list of parents and guardians.

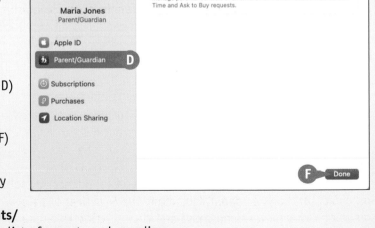

Remove a Person from Family Sharing

To remove a family member aged 13 or older from Family Sharing, click the family member in the Family pane. The family member's configuration dialog opens, with the Apple ID pane displayed. Click **Remove *Name* from Family** (G).

In the dialog that opens next, click **Remove *Person*** (H). In the Remove *Person* from Family? dialog that appears next, click **Remove *Person***.

Once you have added a child younger than 13 to Family Sharing, you cannot remove them from your family except by transferring them to another family. To transfer a child, the organizer of another family group must request the transfer. You, the organizer of your family group, receive a notification of the requested transfer, which you then approve to effect the transfer.

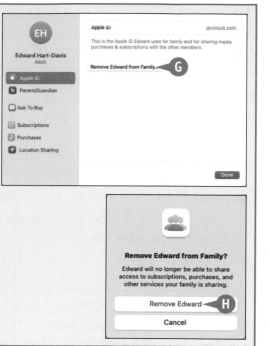

Set Up Location Sharing

To help your family members keep track of each other, you can enable location sharing. In the Family pane, click **Location Sharing** (◢) to open the Location Sharing dialog. In the Share Your Location With box, set the switch for each family member to On (●○, I) or Off (○, J), as needed. In the lower box, set the **Automatically Share Location** switch (K) to On (●○) if you want to share your location automatically with any family members you add to the family later. Click **Done** (L) to close the Location Sharing dialog.

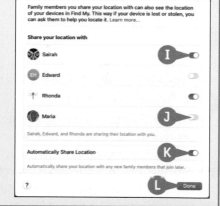

Approve a Purchase

When a family member who does not have authority to make purchases tries to buy an item on a Mac, an iPhone, or an iPad, the app involved prompts them to ask permission to make the purchase. If the family member clicks **Ask**, iCloud sends a notification to the organizer and to each parent/guardian. The organizer or parent/guardian can review the purchase and click **Decline** to refuse it, click **Buy** to authorize payment for an item, or click **Get** to authorize a free item.

Turn On Screen Time for a Child

The Screen Time feature in macOS, iOS, and iPadOS enables you to track how much time a person spends using their devices. You can use Screen Time to monitor your own usage and set limits on it — such as enforcing "downtime," as discussed in the following section, "Configure Downtime" — but you may find it more useful for monitoring and setting limits on what other family members do.

Once you have enabled Family Sharing, you can manage Screen Time through Family Sharing using either the same Mac, iPad, or iPhone that a family member uses or a different Mac or device.

Turn On Screen Time for a Child

1 Press `Control` + click **System Settings** (⚙) on the Dock.

The contextual menu opens.

2 Click **Screen Time**.

The System Settings app opens and displays the Screen Time pane.

3 Click the **Family Member** pop-up menu (⇕), and then click the family member for whom you want to turn on Screen Time.

4 Click **Set Up Screen Time for Your Child**.

The Screen Time dialog opens.

5 Click **Continue**.

The What Content Can Your Child Access? dialog opens.

6 Drag the **Age** slider to set a suitable age range.

7 Click the **Web Content** pop-up menu (⇕), and then click **Unrestricted**, **Limit Adult Websites**, or **Allowed Websites**, as appropriate.

8 Click the **Apps** pop-up menu (⇕), and then click a rating, such as **12+** or **17+**.

9 Scroll down the dialog to display further controls.

84

10. Click the **Movies** pop-up menu (⌃), and then click a rating, such as **PG-13** or **NC-17**.
11. Click the **TV Shows** pop-up menu (⌃), and then click a rating, such as **TV-G**.
12. Click the **Books** pop-up menu (⌃), and then click **Clean** or **Explicit**.
13. Click the **Apple Media** pop-up menu (⌃), and then click **Clean** or **Explicit**.
14. Click the **Music Videos** pop-up menu (⌃), and then click **On** or **Off**.
15. Click the **Music Profiles** pop-up menu (⌃), and then click **On** or **Off**.
16. Click the **Siri Web Search** pop-up menu (⌃), and then click **Allow** or **Don't Allow**.
17. Click the **Explicit Language** pop-up menu (⌃), and then click **Allow** or **Don't Allow**.
18. Click the **Deleting Apps** pop-up menu (⌃), and then click **Allow** or **Don't Allow**.
19. Click **Turn On Restrictions**.

    The Sensitive Photos and Videos Protection dialog opens.
20. Click **Turn On Communication Safety**.

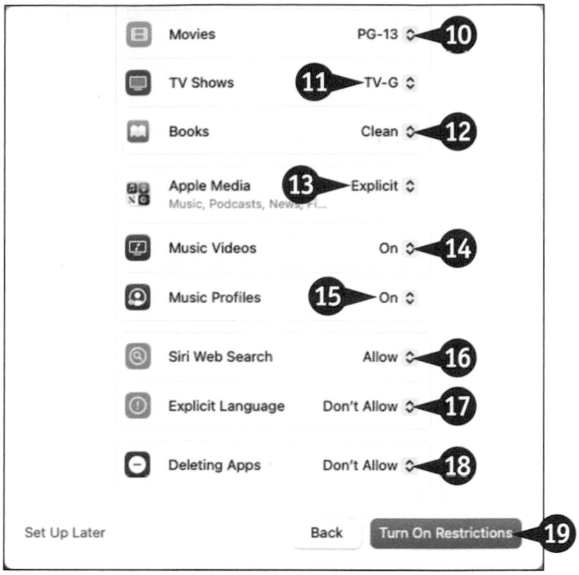

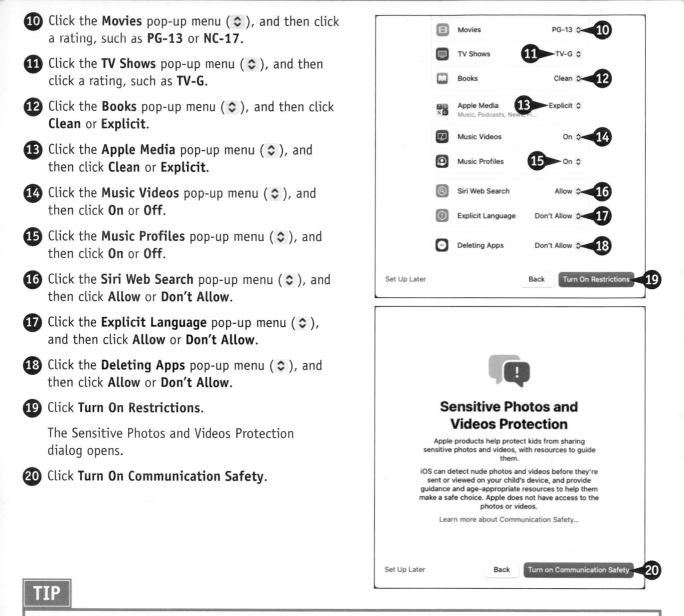

TIP

Should I turn on the Communication Safety feature for my child's account?

Turning on Communication Safety for a child's account is usually a good idea. Communication Safety aims to detect photos and videos containing nudity. When the feature detects potentially problematic photos or videos, it blurs them in apps such as Messages, warns the user that the photos or videos may contain sensitive content, and gives the user ways to get help.

continued ►

Screen Time includes the Screen Distance feature, which encourages the user to keep an iPhone or iPad at a distance of 12 inches and blocks usage at shorter distances. Screen Time also enables you to choose between allowing the user unrestricted access to websites, limiting their access to adult websites, and permitting access only to a list of websites you have approved.

Turn On Screen Time for a Child's Account (continued)

The Teach Your Child to Use Screen Distance dialog opens.

 Read the information.

22 Click **Continue** if you want to use Screen Distance.

A Click **Set Up Later** if you do not want to set up Screen Distance now.

The Keep Track of Your Child's Screen Time? dialog opens.

23 Read the information.

24 Click **Turn On App & Website Activity** if you want to enable Screen Time.

B Click **Set Up Later** if you do not want to set up Screen Time now.

The Set Time Away from Screens? dialog opens.

25 In the **Start** box, enter the start time for downtime, such as **10:00 PM**.

26 In the **End** box, enter the end time for downtime, such as **7:00 AM**.

27 Click **Set Downtime**.

C Click **Set Up Later** if you do not want to set up downtime now.

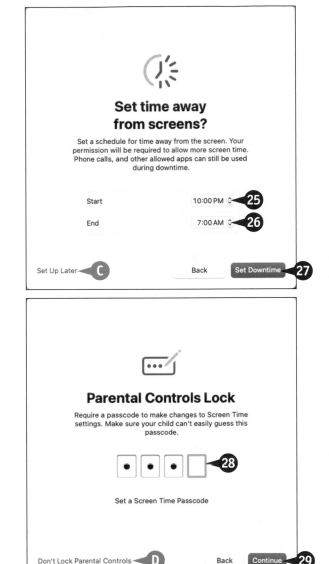

The Parental Controls Lock dialog opens.

28 Type a four-digit passcode, and then type it again when prompted.

29 Click **Continue**.

D Click **Don't Lock Parental Controls** if you prefer not to lock the settings.

The Parental Controls Lock dialog closes.

The Screen Time controls appear in the Screen Time pane.

You can now configure the Screen Time features, as explained in the following sections.

TIP

Should I set a passcode to lock the Parental Controls settings?

Yes, you should set a passcode, because the parental controls are worthless if your child can override them at will. Just make sure you do not forget the passcode.

Configure Downtime

Screen Time's Downtime feature enables you to specify times when the user is not allowed to use a Mac, iPhone, or iPad. For example, you might set downtime from 10:00 PM to 7:00 AM to prevent the user from using a Mac or device during those hours. You can either set the same downtime hours for every day or customize the hours for different days.

Configure Downtime

1 Press **Control**+click **System Settings** (⚙) on the Dock.

The contextual menu opens.

2 Click **Screen Time**.

The System Settings app opens and displays the Screen Time pane at the front.

3 Click the **Family Member** pop-up menu (⌄) and then click the family member for whom you want to configure downtime.

4 Click **Downtime** (⏲).

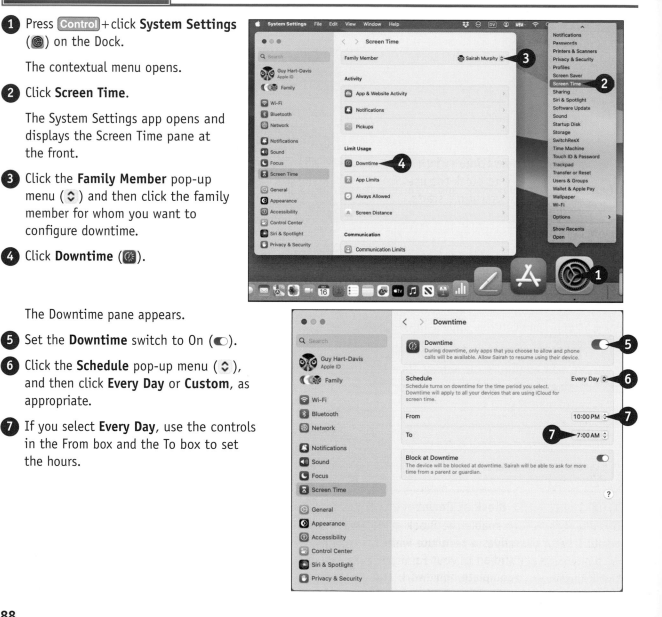

The Downtime pane appears.

5 Set the **Downtime** switch to On (⬤).

6 Click the **Schedule** pop-up menu (⌄), and then click **Every Day** or **Custom**, as appropriate.

7 If you select **Every Day**, use the controls in the From box and the To box to set the hours.

8 If you select **Custom** in the Schedule pop-up menu, use the individual days' controls to specify the downtime schedule.

A If you want to specify no downtime for a particular day, set its switch to Off ( ).

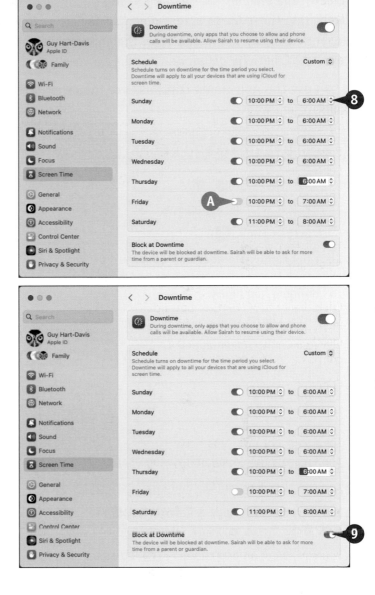

9 Set the **Block at Downtime** switch to On ( ) if you want to block the user from using devices when downtime starts.

TIP

Should I enable the Block at Downtime setting?

Normally, it is best to enable the Block at Downtime setting if you want to enforce the downtime schedule you set. The user receives a 5-minute warning of impending downtime and can then request more time, which any parent/guardian in your Family Sharing group can grant. For example, you might allow a child an extra 30 minutes to complete homework.

Set Time Limits for Apps and Websites

Screen Time enables the Family Sharing organizer to set time limits for a child's use of apps and websites. These limits apply to all the devices with which the child uses their Apple ID for Screen Time. For example, you might allow the child only a short time for games but allow a longer time for apps in the Creativity category.

You can use the App Limits controls, like the other Screen Time controls, to set limits for your own account if you want.

Set Time Limits for Apps and Websites

1. Click **System Settings** (⚙).

    The System Settings app opens.

2. Click **Screen Time** (⧖).

3. Click the **Family Member** pop-up menu (⇅) and then click the family member for whom you want to configure app limits.

4. Click **App Limits** (⧖).

    The App Limits screen appears.

5. Set the **App Limits** switch to On (⬤).

6. Click **Add Limit**.

    The Create a New App Limit dialog opens.

    Ⓐ You can click **Search** (Q) and type a search term.

    Ⓑ You can select **All Apps & Categories** (☑) to set time limits for all apps and categories in one move.

    Ⓒ You can select an app category (☑) to affect all apps in the category.

7. Click **Expand** (> changes to ⌄).

    The app category expands.

8. Select the check box (☑) for the app.

9. In the Time area, either click **Every Day** (◯ changes to ⬤) and click the spin buttons (⇅) to set the same time limit for each day, or click **Custom** (◯ changes to ⬤) to customize the schedule.

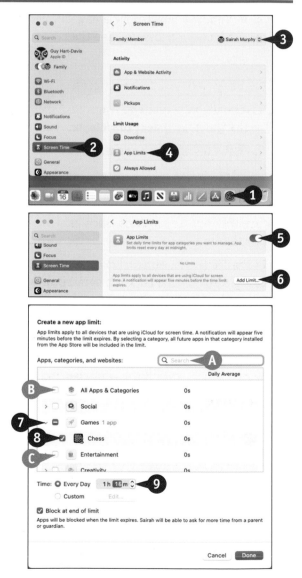

90

Note: Screen Time records the length of time an app is open, regardless of whether the app is being actively used. This blunt measurement method makes it difficult to set effective time limits, because users typically run multiple apps at once and switch among them as needed.

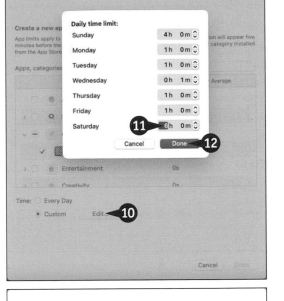

 If you clicked **Custom**, click **Edit**.

The Daily Time Limit dialog opens.

 Set the time limit for each day.

Note: The Daily Time Limit dialog does not let you set a limit of 0 hours 0 minutes for a day, but you can set 0 hours 1 minute, if necessary. See the tip for details.

⓬ Click **Done**.

The Daily Time Limit dialog closes.

⓭ To set daily time limits for websites, click **Expand** (> changes to ⌄) next to Websites.

The Websites category opens.

⓮ Click **Add Website** (⊕).

An edit box opens.

⓯ Type the website's address, such as www.wiley. com, and press Return.

⓰ In the Time area, either click **Every Day** (○ changes to ●) and click the spin buttons (⟳) to set the same time limit for each day, or click **Custom** (○ changes to ●) and click **Edit** to create a custom schedule.

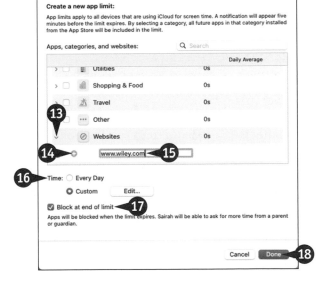

⓱ Select **Block at end of limit** (☑) if you want to block the apps when the user reaches the time limit.

⓲ Click **Done**.

TIP

Why is the Done button grayed out in the Create a New App Limit dialog?
You may have inadvertently set a zero-length time for a day in a custom schedule. After choosing items in the Apps, Categories, and Websites box, click **Custom** (○ changes to ●), click **Edit**, and set a limit for each individual day. Every day must have a limit above 0 hours 0 minutes; if any day has a 0 hours 0 minutes value, the Done button is grayed out.

Make Apps Always Available to a User

The Always Allowed feature in Screen Time enables you to make specific apps always available for a user to use even during downtime. You can use Always Allowed to make sure the user always has those communications apps and features they need to stay in contact and stay safe. You may also want to provide productivity and creativity apps for the user to use when they are blocked from entertainment apps, social networking, and other time sinks.

Make Apps Always Available to a User

1. Click **System Settings** (⚙️).

2. Click **Screen Time** (⏳).

3. Click the **Family Member** pop-up menu (⬍) and then click the family member for whom you want to make apps always available.

4. Click **Always Allowed** (✅).

The Always Allowed screen appears.

5. In the During Downtime box, click **Specific Contacts** (○ changes to ●) or **Everyone** (○ changes to ●).

6. If you choose Specific Contacts, click **Edit Contacts**.

A dialog opens.

7. Check that the list of contacts is appropriate and complete.

Ⓐ You can click **Info** (ⓘ) to display and edit that contact's information.

Ⓑ You can click **Add** (➕) to start adding a contact.

Ⓒ You can select a contact and click **Remove** (➖) to remove them.

8. Click **Done**.

92

The dialog closes.

Ⓓ You can click **Search** (🔍) and type a search term to find an app quickly.

❾ To make the list of apps easier to navigate, click **Sort By** (☰).

The Sort By pop-up menu opens.

❿ Click **Daily Usage**, **Name**, or **State** — whichever you will find better.

Note: Selecting **State** sorts the apps by whether they are allowed or not, putting the allowed apps at the top of the list. State sort is most useful for seeing which apps you've allowed.

Screen Time sorts the list the way you chose.

⓫ Set the switch to On (🔘) for each app you want to make always available to the user.

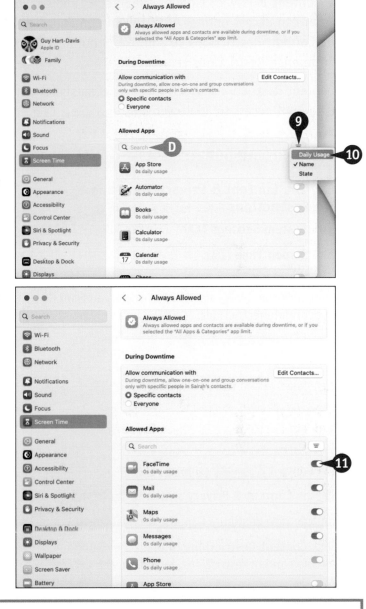

Which apps should I always allow?

At a minimum, you should always allow Phone, Messages, Mail, and Contacts; Maps and FaceTime may be helpful, too. Screen Time prevents you from disallowing Phone to avoid you inadvertently rendering your child incommunicado. Messages can be a vital communications conduit when Phone is not suitable. Similarly, you should make the Mail app always available in case of important messages, and keep Contacts available so that your child can stay in touch at any time. If your child may need to navigate, include Maps as well.

Apply Content and Privacy Restrictions

The Content & Privacy category in Screen Time enables you to choose settings to limit the content a user can access online. After turning on Content & Privacy Restrictions, you can choose settings on the Content Restrictions screen, the Store Restrictions screen, the App Restrictions screen, and the Preference Restrictions screen.

The Content Restrictions screen includes controls for limiting access to web content. You can either set Safari to limit access to adult websites or allow a user access to only the websites on an approved list.

Apply Content and Privacy Restrictions

Display the Content & Privacy Pane and Enable Restrictions

1 Click **System Settings** (⚙).

2 Click **Screen Time** (⧖).

3 Click the **Family Member** pop-up menu (⇕) and then click the family member for whom you want to configure restrictions.

4 Click **Content & Privacy** (⊘).

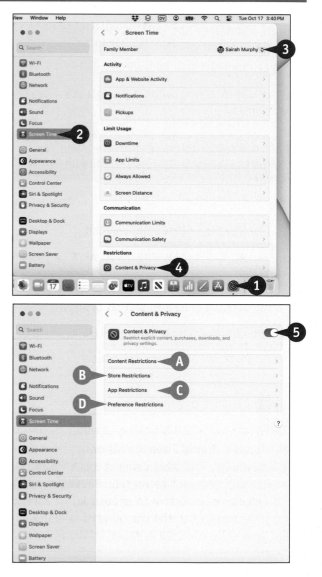

The Content & Privacy pane appears.

5 Set the **Content & Privacy** switch to On (⬤).

You can now configure Content & Privacy restrictions.

Ⓐ Click **Content Restrictions** to open the dialog for setting content restrictions, including web access.

Ⓑ Click **Store Restrictions** to open the dialog for restricting access to the iTunes Store and App Store.

Ⓒ Click **App Restrictions** to open the dialog for restricting what apps and features the user can run.

Ⓓ Click **Preference Restrictions** to open the dialog for restricting changes to key settings, such as passcodes and accounts.

Set Content Restrictions

1 On the Content & Privacy screen, click **Content Restrictions** (not shown).

The dialog for setting content restrictions opens.

2 Click the **Access to Web Content** pop-up menu (⇕), and then click **Unrestricted Access**, **Limit Adult Websites**, or **Allowed Websites Only**.

This example uses Allowed Websites Only, which is the best choice for a child.

Note: When you permit a user to visit only certain websites, those sites appear on the Bookmarks bar in Safari.

3 Click **Customize**.

A dialog opens, showing a prepopulated list of child-friendly websites.

 You can click **Remove** (—) to remove the selected website.

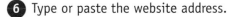 You can click **Settings** (⊙) to change the display name or address for the selected website.

4 Click **Add** (+).

A dialog opens for adding a website.

5 Type the name under which you want to list the website.

6 Type or paste the website address.

7 Click **Done** to close the second dialog.

Note: You can now add other websites, as needed.

8 Click **Done** to close the first dialog.

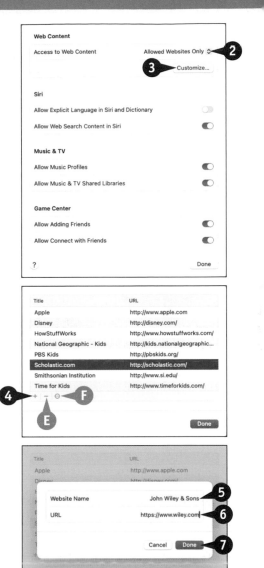

TIP

How effective is the blocking of adult websites?
The blocking of adult websites is only partly effective. macOS, iOS, and iPadOS can block sites that identify themselves as adult sites using standard rating criteria, but many adult sites either do not use ratings or do not rate their content accurately. Because of this, do not rely on Screen Time to block all adult material. It is much more effective to choose **Allow access to only these websites** and provide a list of permitted sites. You can add to the list by vetting and approving extra sites when the user needs to access them.

continued ▶

A part from controlling website access, the Content Restrictions feature also enables you to choose whether to allow explicit language in Siri and the Dictionary app, web search content in Siri, music profiles — artist information — in the Music app, multiplayer games in Game Center, and the ability to add friends in Game Center.

The Store Restrictions feature lets you choose ratings limits for movies, TV shows, and apps; restrict explicit books, music, podcasts, and news; set app restrictions on iOS and iPadOS; and require a password for every purchase on Apple's stores and services.

Apply Content and Privacy Restrictions (continued)

Specify Allowed and Blocked Websites

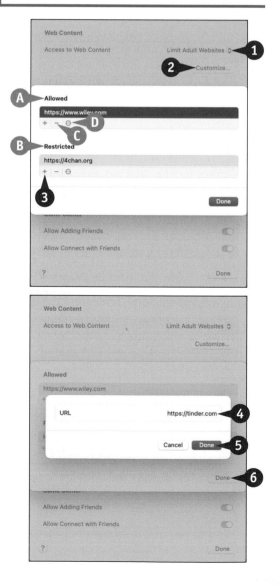

1 In the dialog for setting content restrictions, click the **Access to Web Content** pop-up menu (⬍), and then click **Limit Adult Websites**.

2 Click **Customize**.

A dialog opens.

Ⓐ The Allowed list box shows allowed websites.

Ⓑ The Restricted list box shows restricted — blocked — websites.

Ⓒ You can click **Remove** (—) to remove the selected website.

Ⓓ You can click **Settings** (⊙) to change the address for the selected website.

3 Click **Add** (+) below the Allow list box or the Restricted list box. This example uses Restricted.

A dialog opens for adding a website.

4 Type or paste the website address.

Note: Although the setting on the Content tab of Content & Privacy settings is called Limit Adult Websites, you can put any websites you want on the Allow list and the Restricted list.

5 Click **Done**.

The dialog closes.

Screen Time adds the website to the appropriate list.

6 When you finish adding websites to the lists, click **Done**.

The dialog closes.

Choose Which Content to Allow

1 In the dialog for setting content restrictions, set the **Allow Explicit Language in Siri and Dictionary** switch to On (⬤) or Off (◯), as needed.

2 Set the **Allow Web Search Content in Siri** switch to On (⬤) or Off (◯), as needed.

3 Set the **Allow Music Profiles** switch to On (⬤) or Off (◯), as needed.

4 Set the **Allow Music & TV Shared Libraries** switch to On (⬤) or Off (◯), as needed.

5 Set the **Allow Adding Friends** switch to On (⬤) or Off (◯), as needed.

6 Set the **Allow Connect with Friends** switch to On (⬤) or Off (◯), as needed.

7 Scroll down.

8 Set the **Allow Private Messaging** switch to On (⬤) or Off (◯), as needed.

9 Set the **Allow Avatar & Nickname Changes** switch to On (⬤) or Off (◯), as needed.

10 Set the **Allow Profile Privacy Changes** switch to On (⬤) or Off (◯), as needed.

11 Click the **Allow Multiplayer Games With** pop-up menu (⬍), and then click **Everyone**, **Friends Only**, or **No One**.

12 Set the **Allow Nearby Multiplayer** switch to On (⬤) or Off (◯), as needed.

13 Click **Done** to close the dialog.

Web Content

| Access to Web Content | Limit Adult Websites ⬍ | **7** |

Customize...

Siri

Allow Explicit Language in Siri and Dictionary ◯ **1**
Allow Web Search Content in Siri ⬤ **2**

Music & TV

Allow Music Profiles ⬤ **3**
Allow Music & TV Shared Libraries ⬤ **4**

Game Center

Allow Adding Friends ⬤ **5**
Allow Connect with Friends ⬤ **6**

? Done

Music & TV

Allow Music Profiles ⬤
Allow Music & TV Shared Libraries ⬤

Game Center

Allow Adding Friends ⬤
Allow Connect with Friends ⬤
Allow Private Messaging ⬤ **8**
Allow Avatar & Nickname Changes ⬤ **9**
Allow Profile Privacy Changes ⬤ **10**

Multiplayer Games

Allow Multiplayer Games With Everyone ⬍ **11**
Allow Nearby Multiplayer ⬤ **12**

? **13** Done

TIP

Which Require Password setting should I choose?

For a child, choose **Always Require** (◯ changes to ⦿) to make your Macs, iPads, and iPhones require a password for every single purchase from Apple's services. Otherwise, authorizing a single purchase gives the child 15 minutes to purchase other apps or items with few restrictions. Free-to-play games marketed to children often include in-app purchases that can rack up costs quickly, especially if implemented in a predatory way.

continued ▶

The Store Restrictions feature in Content & Privacy settings enables you to control what content the user can access on the App Store and the iTunes Store. The App Restrictions feature lets you specify what apps the user is allowed to run. The Preference Restrictions feature controls whether the user can configure various key features on iOS and iPadOS devices for themselves. For example, you can prevent the user from changing their passcode; altering Cellular Data settings, such as Data Roaming; or reconfiguring the Driving Focus settings.

Apply Content and Privacy Restrictions (continued)

Set Store Restrictions

1 On the Content & Privacy screen, click **Store Restrictions** (not shown).

The dialog for setting store restrictions opens.

2 Click **Movies** (↕), and then click the highest certification to allow, such as **PG-13** or **R**.

Note: To block movies, TV shows, or apps completely, click the appropriate pop-up menu (↕), and then click **Don't Allow Movies**, **Don't Allow TV Shows**, or **Don't Allow Apps**.

3 Click **TV Shows** (↕) and then click the highest certification to allow, such as **TV-PG** or **TV-14**.

4 Click **Apps** (↕) and then click the highest age rating to allow, such as **12+**.

5 Set the **Allow Explicit Books** switch, the **Allow Explicit Music, Podcasts, and News** switch, and the **Allow Music Videos** switch to On (●) or Off ( ), as needed.

6 Set the **Allow Installing Apps** switch to On (●) or Off ( ), as needed.

7 Set the **Allow Deleting Apps** switch to On (●) or Off ( ), as needed.

8 Set the **Allow In-app Purchases** switch to On (●) or Off ( ), as needed.

9 Click the **Require Password** pop-up menu (↕), and then click **Always** or **After 15 Minutes**, as needed.

10 Click **Done** to close the dialog.

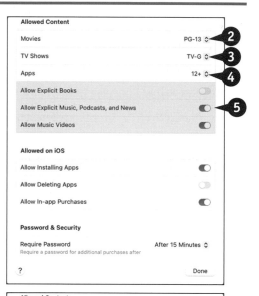

Set App Restrictions

1 On the Content & Privacy screen, click **App Restrictions** (not shown).

The dialog for setting app restrictions opens.

2 In the Allowed area, set the **Allow Camera** switch, the **Allow Book Store** switch, the **Allow Siri & Dictation** switch, and the **Allow SharePlay** switch to On () or Off (⬭), as needed.

3 In the Allowed on iOS area, set the switch for each app or feature to On (⬤) or Off (⬭), as needed. Scroll down to reach all the settings.

4 Click **Done** to close the dialog.

Set Preference Restrictions

1 On the Content & Privacy screen, click **Preference Restrictions** (not shown).

The dialog for setting preference restrictions opens.

2 Set the **Allow Passcode Changes** switch to On (⬤) or Off (⬭), as needed.

3 Set the **Allow Account Changes** switch to On (⬤) or Off (⬭), as needed.

4 Set the **Allow Cellular Data Changes** switch to On (⬤) or Off (⬭), as needed.

5 Set the **Allow Driving Focus Changes** switch to On (⬤) or Off (⬭), as needed.

6 Set the **Allow TV Provider Changes** switch to On (⬤) or Off (⬭), as needed.

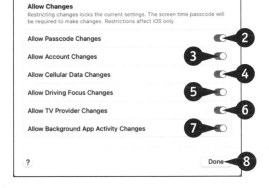

7 Set the **Allow Background App Activity Changes** switch to On (⬤) or Off (⬭), as needed. See the tip for details.

8 Click **Done** to close the dialog.

TIP

What are background app activities?

On an iPhone, the app you are using is considered to be in the foreground; on an iPad, you can have two or more foreground apps at once. Any non-foreground apps visible in the App Switcher are considered to be running in the background; other apps not visible in the App Switcher may also be performing actions, such as tracking your location. iOS and iPadOS enable you to allow apps to perform activities while in the background. For example, Mail synchronizes your e-mail messages with your e-mail providers while in the background, Messages synchronizes your messages, and the Phone app listens for incoming phone calls.

Review a User's Actions

After you, as the family organizer, enable Screen Time for a child user, Screen Time logs the actions that user takes when logged in to devices using their Apple ID. You can review the logs of a user's actions to see what the user has done and what they have tried to do. Using this information, you can decide whether you need to adjust the Screen Time settings to allow more freedom or tighter control. You can review the Screen Time logs either from your MacBook or from another device you log in to using your Apple ID.

Review a User's Actions

1 Click **Apple** (🍎).

The Apple menu opens.

2 Click **System Settings**.

The System Settings window opens.

3 Click **Screen Time** (⬛).

The Screen Time pane appears.

4 Click **Family Member** (↕), and then click the family member whose actions you want to review.

5 Click **App & Website Activity** (📊).

The App & Website Activity screen appears.

6 Navigate to the date you want to view:

Ⓐ You can click **Today** to display the current date.

Ⓑ You can click **Choose Date** (↕), and then click **This Week** to display the week's data.

Ⓒ You can click **Previous** (<) to display the previous day.

Ⓓ You can click **Next** (>) to display the next day.

Ⓔ The user's usage data for the day appears.

Ⓕ This readout shows when the data was last updated.

Ⓖ You can switch between viewing apps and categories by clicking the **Show** pop-up menu (↕) and then clicking **Show Apps** or **Show Categories**.

7 Click **Back** (<).

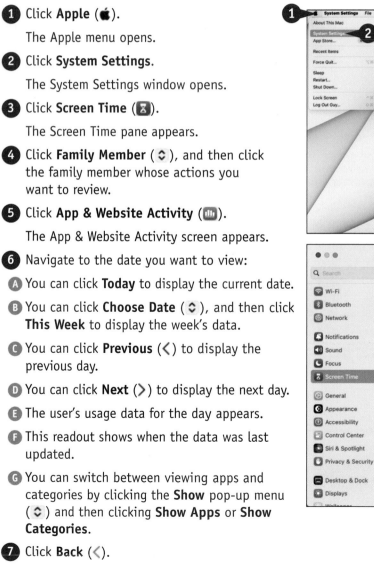

The Screen Time screen appears again.

8 Click **Notifications** (■) (not shown).

The Notifications screen appears.

H You can see how many notifications the user has received.

I You can see when the notifications occurred.

9 Click **Back** (<).

The Screen Time screen appears once more.

10 Click **Pickups** (■) (not shown).

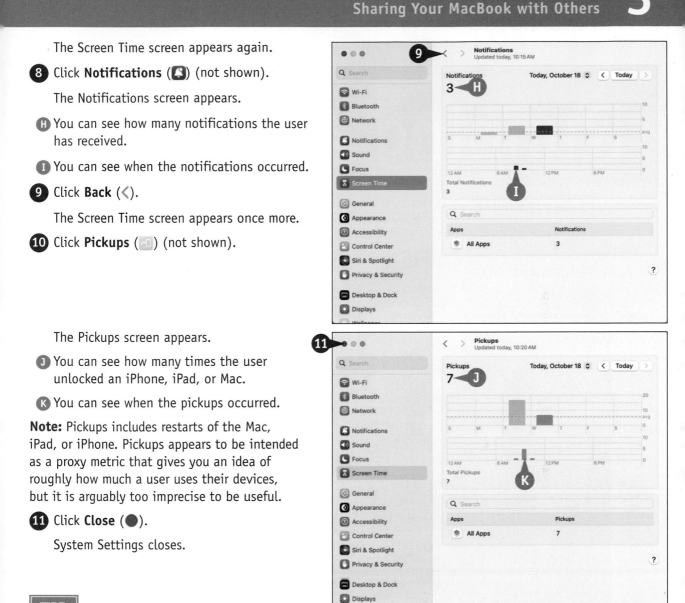

The Pickups screen appears.

J You can see how many times the user unlocked an iPhone, iPad, or Mac.

K You can see when the pickups occurred.

Note: Pickups includes restarts of the Mac, iPad, or iPhone. Pickups appears to be intended as a proxy metric that gives you an idea of roughly how much a user uses their devices, but it is arguably too imprecise to be useful.

11 Click **Close** (●).

System Settings closes.

Does receiving many notifications indicate a problem?

Not necessarily. Various apps raise a large number of notifications under normal circumstances. For example, the Music app raises a notification displaying the details of each song it starts playing, so it may raise around 15–20 notifications per hour if the user is playing popular music.

However, if a child is receiving so many notifications as to distract them from the computer tasks you encourage, you may want to explore what is causing the notifications and consider changing notification settings, or limiting problematic apps or categories of apps, to reduce the onslaught of notifications.

CHAPTER 4

Running Apps

macOS includes many apps, such as the TextEdit word processor, the Preview viewer for PDF files and images, and the Music app for playing music. You can install other apps as needed. Whichever apps you run, you can easily switch among them, quit them when you finish using them, and force them to quit if they crash.

Open an App and Quit It

macOS enables you to open your MacBook's apps in several ways. The Dock is the quickest way to launch apps you use frequently. Launchpad is a handy way to see all the apps installed in your MacBook's Applications folder and its subfolders. You can also launch an app from the Applications folder, but typically you do not need to do so.

When you finish using an app, you quit it by giving a Quit command. You can quit an app either from the menu bar or by using a keyboard shortcut.

Open an App and Quit It

Open an App from the Dock

1 Click the app's icon on the Dock.

Note: If you do not recognize an app's icon, position the pointer over it to display the app's name.

A The app opens.

Open an App from Launchpad

1 Click **Launchpad** (▦) on the Dock.

The Launchpad screen appears.

To scroll to another screen, swipe left or right with two fingers on the trackpad.

B You can also click a dot to move to another screen.

2 Click the app.

The app opens.

Note: To launch an app you have used recently, click **Apple** (), highlight **Recent Items**, and then click the app in the Applications list.

Open an App from the Applications Folder

1 Click **Finder** () on the Dock.

A Finder window opens.

2 Click **Applications** in the left column.

An icon appears for each app.

3 Double-click the app you want to run.

The app opens.

Note: You can also open an app by clicking **Search** (Q), starting to type the app's name, and then clicking the appropriate search result.

Quit an App

1 Click the app's menu, the menu with the app's name — for example, **Dictionary**.

The menu appears.

2 Click the Quit command from the menu that has the app's name — for example, **Quit Dictionary**.

Note: You can also quit the active app by pressing ⌘+Q.

Note: You can quit some single-window apps, such as System Settings and Dictionary, by clicking **Close** (●). But for most apps, clicking **Close** (●) closes the window but leaves the app running

How do I add an app to the Dock?

Open the app as usual — for example, by using Launchpad. Press Control +click its Dock icon, click or highlight **Options** on the contextual menu, and then click **Keep in Dock**. You can also click an app in Launchpad or the Applications folder and drag it to the Dock.

What happens if a document in the app I quit contains unsaved changes?

Some apps automatically save your changes when you quit the app. Other apps display a dialog asking if you want to save the changes. Click **Save** to save the changes, **Revert** to discard the changes and revert to the last saved version of the document, or **Delete** to delete an unsaved document.

Install an App

Your MacBook comes with many useful apps already installed, such as Safari for browsing the Web, Mail for reading and sending e-mail, and Music for enjoying music and video.

To get your work or play done, you may need to install other apps on your MacBook. You can install apps in two main ways: by downloading them from Apple's App Store using the App Store app, or by downloading them from other websites and installing them manually. If you acquire an app on CD or DVD, connect an optical drive to your MacBook via USB, load the disc, and follow the prompts.

Install an App from the App Store

Apple's App Store offers a wide selection of apps built to run on macOS. Apple vets the developers of apps and tests each app they submit for inclusion on the App Store, so you can be sure that each app contains no malicious code and will run on your MacBook without causing problems. Each app also includes reviews and ratings, many of them carefully considered and helpful. Because of the App Store's security precautions and quality control, you should make it your first choice for getting apps for your MacBook.

Click **App Store** (A,) on the Dock to open the App Store app. You can then locate the app you want to install either by clicking in the Search box (B) and typing the app's name or your search terms or by browsing to the app — for example, by clicking the appropriate category in the sidebar (C) on the left.

After locating the app, click **Get** (D) or the price button, enter your Apple ID email address and password if prompted to do so, and then wait while App Store installs the app. You can then run the app by clicking Launchpad ( ) on the Dock, and then clicking the app's icon on the Launchpad screen.

When new versions of apps are available, you should install them so that you can benefit from the latest features added and any fixes for problems. See the section "Keep Your MacBook Current with Updates" in Chapter 12 for information about updating apps.

Download and Install an App from a Website

If the app you want to install is not available on the App Store, you can most likely acquire it as a file that you download from a website and then install manually on your MacBook.

To get such an app, go to the software company's official website; if you do not know the address, search for it. Locate the appropriate link — for example, by following a Products link or a Downloads link — and then click it to download the file to your MacBook's Downloads folder. If macOS prompts you to allow downloads from the website, click **Allow** (E).

Click **Finder** on the Dock to open a Finder window, and then click **Downloads** (⬇) in the sidebar to display the contents of the Downloads folder. Double-click the downloaded file to open it.

If the app uses an installer, macOS displays a dialog to verify that you want to open the app. Make sure that the message *Apple checked it for malicious software and none was detected* (F) appears, and then click **Open** (G). Follow the installer's prompts to choose any options available for the installation and to complete the installation.

If the app installs as a single package, a Finder window opens. Drag the app's icon (H) to the Applications folder (I) to install the app. Click **Close** (J, ⬤) to close the Finder window, press Control+click or right-click the app's disk icon on the Desktop, and then click **Eject**. If macOS then prompts you to delete the package file, either accept the offer or delete it later from the Downloads folder at your convenience.

You can now run the installed app from Launchpad.

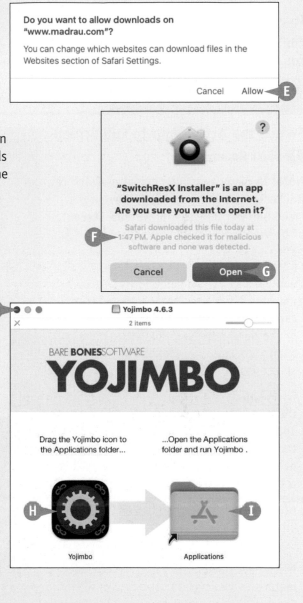

Run an App Full Screen

macOS enables you to run an app full screen instead of in a window. Running an app full screen helps you focus on that app, removing the distraction of other open apps.

You can instantly switch the active app to full-screen display. When you need to use another app, you can switch to that app full screen as well — and then switch back to the previous app. When you finish using full-screen display, you can switch back to displaying the app in a window.

Run an App Full Screen

Switch the Active App to Full Screen

1 Click **Zoom** (●).

Note: In many apps, you can also switch to full screen either by clicking **View** on the menu bar and selecting **Enter Full Screen** or by pressing Control+⌘+F or 🌐+F.

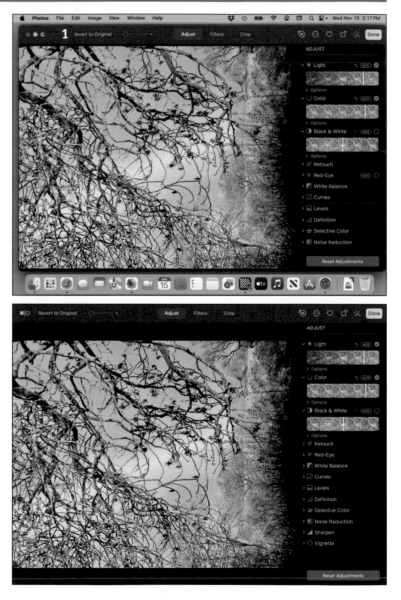

The app expands to take up the full screen.

Switch to Another App

1 Swipe left or right with three fingers on the trackpad.

Note: You can also switch apps by using Application Switcher or Mission Control.

The next app or previous app appears.

2 Swipe in the opposite direction (not shown).

The app you were using before you switched appears.

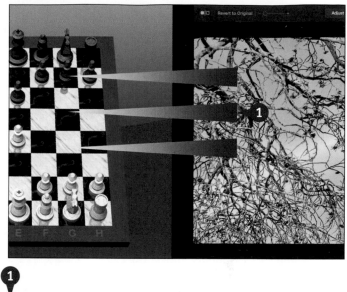

Return from Full-Screen Display to a Window

1 Move the pointer to the upper-left corner of the screen.

The menu bar and the app's title bar appears.

2 Hold the pointer over Zoom (●).

Zoom (●) changes to Zoom Back (●).

The pop-up menu appears.

3 Click **Exit Full Screen** (■).

The app appears in a window again.

Note: In many apps, you can also move the pointer to the top of the screen, click **View**, and then click **Exit Full Screen** on the View menu.

TIPS

How do I display the Dock when an app is in full-screen view?

Move the pointer to the bottom of the screen; the Dock slides into view. If you have positioned the Dock at the left side or right side of the screen, move the pointer to that side to display the Dock.

Can I exit full-screen display by using the keyboard?

Yes. Press Esc once or twice to return from full-screen view to windowed view. In many apps, you can also press ⌘ + Control + F or ⊕ + F to exit full-screen display.

Set Apps to Run Automatically at Login

macOS enables you to set apps to open automatically each time you log in to your MacBook. By opening your most-used apps automatically, you can save time getting started with your work or play. Opening apps at login both makes the login process take longer and increases use of memory and other system resources, so it is best to run only those apps you always use. You can configure an app to open automatically either from the Dock or by using the Login Items pane in System Settings.

Set Apps to Run Automatically at Login

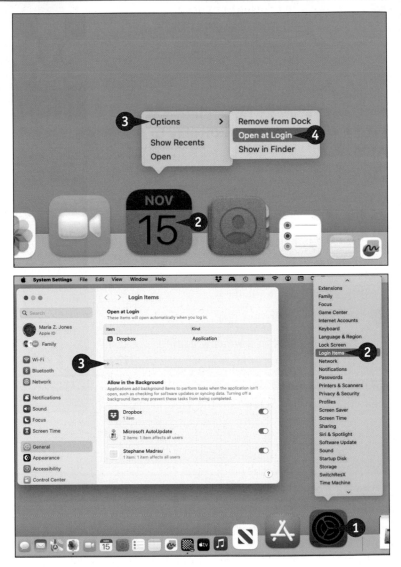

Using the Dock to Set an App to Run at Login

1 If the app does not have a Dock icon, click **Launchpad** ( ) on the Dock and then click the app (not shown).

The app's icon appears on the Dock.

2 Press Control +click the app's Dock icon.

The contextual menu opens.

3 Click or highlight **Options**.

The Options continuation menu opens.

4 Click **Open at Login**.

A check mark appears next to Open at Login.

Using System Settings to Set an App to Run at Login

1 Press Control +click **System Settings** ( ) on the Dock.

The contextual menu opens.

Note: If System Settings is already open, click **General** ( ) to display the General pane, and then click **Login Items** ( ).

2 Click **Login Items**.

The Login Items pane appears.

3 Click **Add** (+).

A dialog opens.

4 Navigate to the appropriate folder. For example, click **Applications** (🅰) to display the apps in the Applications folder.

Note: You can also add documents to the list for automatic opening. To do so, navigate to the document, click it, and then click **Open**. macOS opens the document in its default app at login.

5 Click the app you want to run automatically at login.

Note: To select multiple apps, click the first, and then press ⌘+click each of the others.

6 Click **Open**.

The dialog closes.

Ⓐ The app appears in the list.

7 Click **Close** (●).

System Settings closes.

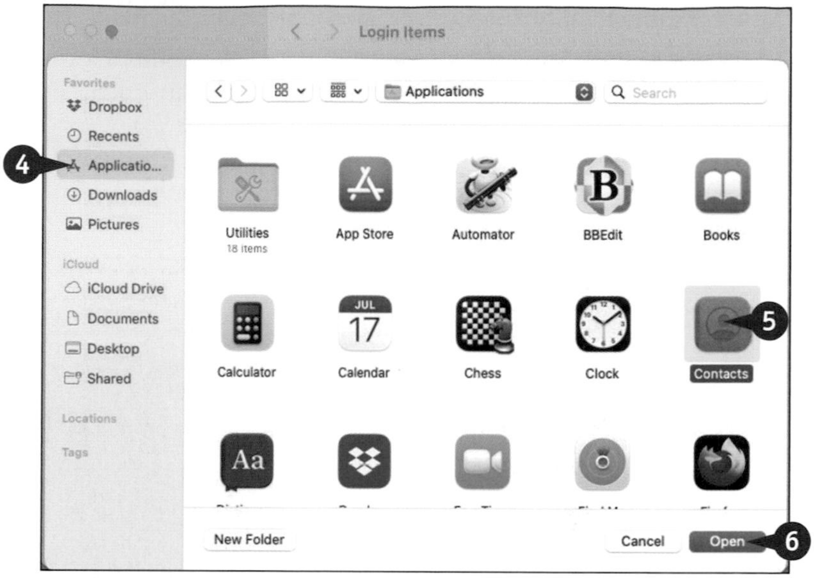

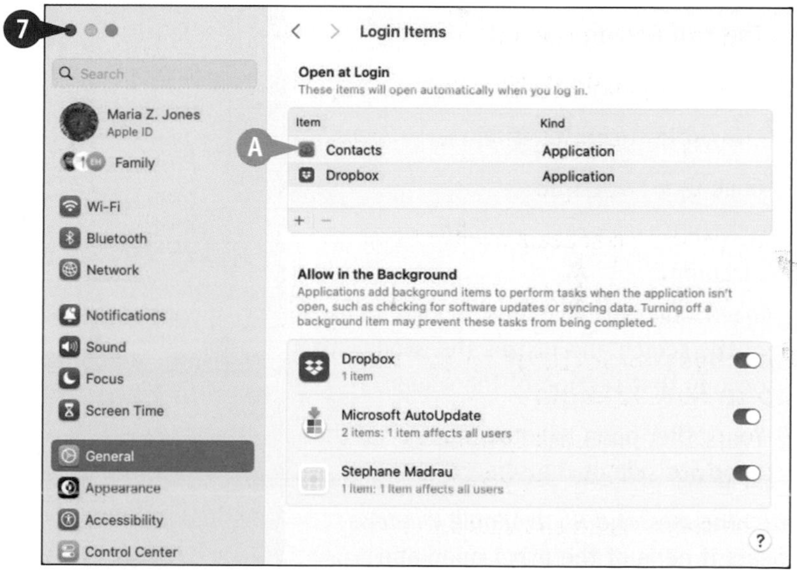

TIP

How else can I add an app to the Login Items pane?

Instead of clicking **Add** (+) and using the dialog to pick the apps, you can drag the apps from a Finder window. Click **Finder** (🙂) on the Dock to open a Finder window, and then click **Applications** (🅰) in the sidebar. Click the app you want, and then drag it across to the Login Items pane in the System Settings window.

Using Split View

macOS includes Split View, which enables you to divide the screen between two apps. When you need to view two apps simultaneously, using Split View can save time and effort over resizing and positioning app windows manually.

Depending on the apps you choose and the screen resolution on your MacBook, the split in Split View may not be equal. As of this writing, only some apps work in Split View, but macOS clearly identifies app windows that are not available in Split View.

Using Split View

1 Hold the pointer over **Zoom** (●) in the window of the first app you want to use in Split View.

The pop-up menu opens.

2 Click **Tile Window to Left of Screen** (■) or **Tile Window to Right of Screen** (■), as needed.

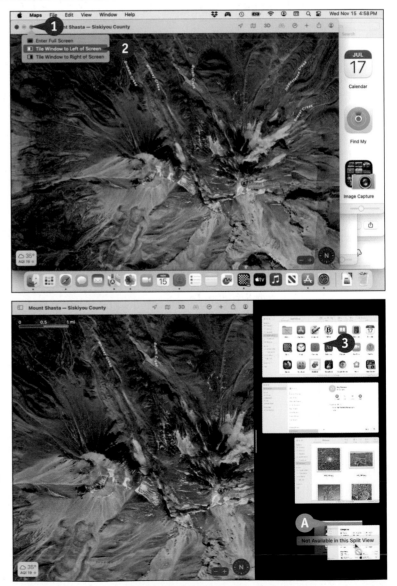

macOS snaps the window to that side of the screen and resizes the window to occupy that section of the screen.

Your other open windows appear on the opposite side of the screen.

Note: The message *No Available Windows* appears if none of the other open app windows is available in Split View.

Ⓐ A small thumbnail indicates that the window is not available in this split view. If you move the pointer over such a window, the message *Not Available in This Split View* appears.

3 Click the window you want to position on the other section of the screen.

macOS snaps the window to the side of the screen and resizes the window to occupy the other section of the screen.

You can then work in the app windows as normal.

B You can resize the windows by dragging the handle on the separator bar between them.

4 When you are ready to finish using Split View, move the pointer to the upper-left corner of a window.

The title bar for each window appears.

C You can click **Zoom Back** ( ) to restore the windows to their previous sizes and positions.

D Alternatively, you can hold the pointer over Zoom Back until the pop-up menu opens.

E You can click **Replace Tiled Window** ( ) to choose another window to replace this one.

F You can click **Move Window to Desktop** ( ) to restore the window to its previous size and make it active.

G You can click **Make Window Full Screen** ( ) to make the window full screen.

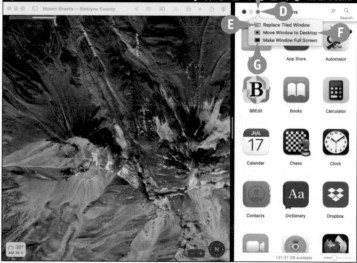

TIPS

How do I move an app to the other side in Split View?

Move the pointer to the top of the app's window. When the title bar appears, click it and drag it to the other side of the screen. macOS switches the two windows.

Are there other ways to switch to Split View?

If you are using an app full screen, swipe up with three fingers on the trackpad to open Mission Control. You can then drag a window onto the full-screen app's thumbnail in the Spaces bar at the top of the screen. macOS switches the apps to Split View on the thumbnail. Click the space's thumbnail to display that space full screen.

Switch Quickly Among Apps

macOS enables you to switch quickly among your open apps by using either the trackpad or the keyboard.

If you have several apps displayed on-screen, you may be able to switch by clicking the window for the app you want to use. If the app is not visible, you can click the app's icon on the Dock. If the app has multiple windows, you can then select the window you need. Alternatively, you can use the App Switcher feature to move quickly from one open app to another.

Switch Quickly Among Apps

Switch Apps Using the Trackpad

1. If you can see a window for the app to which you want to switch, click anywhere in that window.

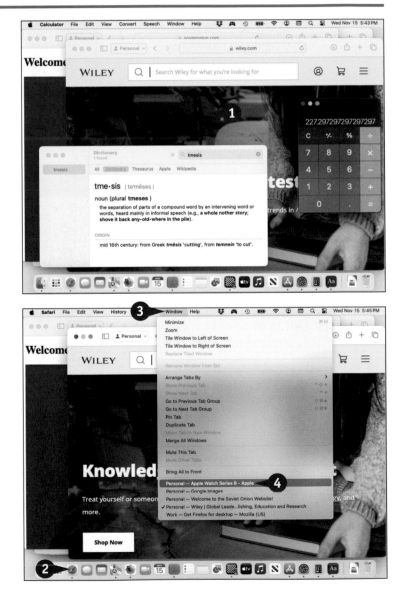

2. If you cannot see a window to which you want to switch, click the app's icon on the Dock.

   All the windows for that app appear in front of the other apps' windows.

3. Click **Window** on the menu bar.

   The Window menu opens.

4. Click the window you want to bring to the front.

Note: To bring a specific window to the front, press Control +click the app's icon on the Dock to display the contextual menu, and then click the window you want to see. You can also click and hold the app icon on the Dock to display the contextual menu.

Switch Among Apps Using the Keyboard

1 Press and hold ⌘ and press Tab (not shown).

A App Switcher opens, showing an icon for each open app.

2 Still holding down ⌘, press Tab one or more times to move the highlight to the app you want (not shown).

Note: Press and hold ⌘+Shift and press Tab to move backward through the apps.

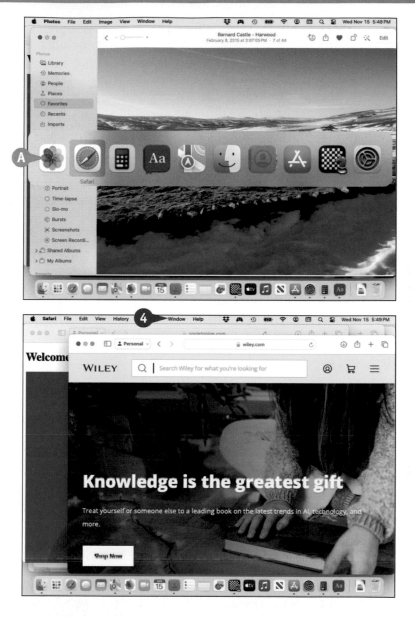

3 When you reach the app you want, release ⌘ (not shown).

App Switcher closes, and the selected app comes to the front.

4 If necessary, click **Window** and select the window you want.

Note: You can press ⌘+`~` to switch among the windows in the current app.

TIPS

Are there other ways of switching among apps?

You can use the keyboard and the trackpad — or mouse, if you have connected one — together. Press and hold ⌘, press Tab once to open Application Switcher, and then click the app you want to bring to the front.

Can I do anything else with App Switcher other than switching to an app?

You can also hide an app or quit an application from App Switcher. Press and hold ⌘, press Tab to open App Switcher, and then press Tab as many times as needed to select the app you want to affect. Still holding ⌘, press H to hide the app or press Q to quit the app.

Switch Apps Using Mission Control

The Mission Control feature helps you manage your desktop and switch among apps and windows. When you activate Mission Control, it shrinks the open windows so that you can see them all and click the one you want. You can use Mission Control to display all open windows in all apps or just the windows in a particular app.

Mission Control also shows different desktop spaces, enabling you to switch among desktop spaces or move an app window from one desktop space to another.

Switch Apps Using Mission Control

Switch Among All Your Open Apps and Windows

1 Swipe up on the trackpad with three fingers (not shown).

Note: On most MacBook keyboards, you press F3 to launch Mission Control. You can also press Control + ⬆ on any Mac.

Note: You can also launch Mission Control by using a hot corner. See Chapter 2 for instructions on setting and using hot corners.

Mission Control opens and displays all open apps and windows on the current desktop.

Ⓐ The Spaces bar at the top of the window shows the open desktops and any apps that are full screen.

Ⓑ The current desktop or app is highlighted.

2 Move the pointer over the Spaces bar.

Thumbnails of the desktops and full-screen apps appear on the Spaces bar.

Ⓒ To preview a window in an app, position the pointer over the window so that a blue outline appears around it. Then press Spacebar to preview the window. Press Spacebar again to close the preview.

③ Click the window you want to use.

The window appears, and you can work with it.

Switch Among All the Windows in the Active App

① Click the Dock icon of the app you want to see.

② Press Control + ↓ (not shown).

Note: On some MacBooks, you can also press Control + F3 to display the windows of the active app.

macOS displays thumbnails of that app's windows.

③ Click the window you want to see.

macOS restores all the windows from all the apps, placing the window you clicked at the front.

TIP

What other actions can I take with Mission Control?
After pressing Control + ↓ or Control + F3 to show all windows of the current app, you can press Tab to show all windows of the next app. Press Shift + Tab to show all windows of the previous app. You can press ⌘ + F3 to move all open windows to the sides of the screen to reveal the desktop. Press ⌘ + F3 when you want to see the windows again.

Enable and Configure Stage Manager

As well as App Switcher and Mission Control, discussed earlier in this chapter, macOS provides a feature called Stage Manager for switching among the apps you're working with. You can use Stage Manager instead of App Switcher and Mission Control if you like, or you can use all three features together.

This section explains how to enable and configure Stage Manager. The following section shows you how to switch apps using Stage Manager.

Enable and Configure Stage Manager

① Click **System Settings** (⚙) on the Dock.

The System Settings window opens.

② Click **Desktop & Dock** (▣) in the sidebar.

The Desktop & Dock pane appears.

③ Scroll down to the Desktop & Stage Manager section.

④ On the Show Items row, select **In Desktop** (☑) or **In Stage Manager** (☑) to tell macOS where to display desktop items.

⑤ Click the **Click wallpaper to reveal desktop** pop-up menu (⇕), and then click **Always** or **Only in Stage Manager**.

⑥ Set the **Show recent apps in Stage Manager** switch to On (⬤) to make Stage Manager display the Recent Apps list.

⑦ Click the **Show windows from an application** pop-up menu (⇕), and then click **All at Once**. The other option is One at a Time.

Note: You may want to change the Show Recent Apps in Stage Manager setting and the Show Windows from an Application setting after you start to use Stage Manager.

⑧ Click to set the **Stage Manager** switch to On (⬤).

The Stage Manager dialog opens.

Note: The Stage Manager dialog opens only the first time you enable Stage Manager.

9 Read the explanation.

10 Click **Turn On Stage Manager**.

The Stage Manager dialog closes.

The Stage Manager switch changes to On (⬤).

11 Click **Close** (⬤).

System Settings closes.

The app that was active before you opened System Settings becomes active again.

Stage Manager rearranges the windows of the active app and your other recent apps.

Ⓐ The active window appears in the main part of the screen.

Ⓑ The Recent Apps list appears on the left of the screen.

TIP

Can I change the number of apps in the Recent Apps list?
No — or at least, not directly. Stage Manager varies the number of apps in the Recent Apps list to suit the screen resolution your MacBook is using. Four apps appear in the Recent Apps list at low resolution; more appear at higher resolutions. So unless you change the resolution, you cannot change the number of apps in the Recent Apps list.

Switch Apps Using Stage Manager

After configuring Stage Manager, as explained in the previous section, you can use Stage Manager to switch quickly between the active app and your recent apps. Stage Manager places the active window center stage and arranges a handful of the most recently used apps in a stack on the left side of the screen. You can click an app in the stack to make it active, replacing the center window.

If Stage Manager is not currently active, you can enable it quickly from Control Center or from the menu bar.

Switch Apps Using Stage Manager

1 Click **Control Center** (⬛).

Control Center opens.

2 Click **Stage Manager** (⯾ changes to ⯾).

macOS enables Stage Manager.

Ⓐ The active window takes center stage.

Ⓑ The other recent apps appear in a vertical stack in the Recent Apps list.

3 Click **Control Center** (⬛).

Control Center closes.

Note: You can resize and reposition the active window as needed. If you move the active window into the space occupied by the Recent Apps list, Stage Manager hides the Recent Apps list. You can then display the Recent Apps list by moving the pointer to the left side of the screen.

④ Click the app you want to use.

⊙ The app takes center stage.

⊙ The previously active app appears in the Recent Apps list.

Note: You can drag a thumbnail from the Recent Apps list onto the active app window to group the two apps.

Note: If you want to stop using Stage Manager, click **Control Center** (▨) on the menu bar, and then click **Stage Manager** (⊡).

TIP

What is the quickest way to enable Stage Manager?
You can put Stage Manager on the macOS menu bar so that you can enable and disable the feature without needing to open Control Center. Click **System Settings** (⚙) on the Dock to open the System Settings window, and then click **Control Center** (▨) in the sidebar to display the Control Center pane. In the Control Center Modules list, click the **Stage Manager** pop-up menu (⌄), and then click **Show in Menu Bar**.

Set Up Dictation and Spoken Content

The Dictation feature enables you to dictate text, which can be a fast and accurate way of entering text into documents. The Spoken Content feature enables you to have the system voice read on-screen items to you.

Before using Dictation and Spoken Content, you use Dictation settings and Spoken Content settings to enable the features. You can select your dictation language and the system voice, define keyboard shortcuts for starting dictation and having your MacBook speak text, and choose whether to have Spoken Content announce when alerts are displayed or apps need your attention.

Set Up Dictation and Spoken Content

1. Click **System Settings** (⚙) on the Dock to open the System Settings window.

2. Click **Keyboard** (⌨) to display the Keyboard pane.

3. Scroll down to the Dictation section.

4. Click the **Microphone source** pop-up menu (⌄) and then click the appropriate microphone.

5. Click the **Shortcut** pop-up menu (⌄) and then click the shortcut for starting dictation.

6. Set the **Auto-punctuation** switch to On (⬤) if you want Dictation to punctuate automatically.

7. Click to set the **Dictation** switch to On (⬤).

   The Do You Want to Enable Dictation? dialog opens.

8. Click **Enable**.

   The Do You Want to Enable Dictation? dialog closes.

   System Settings enables Dictation.

9. Scroll up the sidebar.

10. Click **Accessibility** (♿) to display the Accessibility pane.

11. Click **Spoken Content** (🗨).

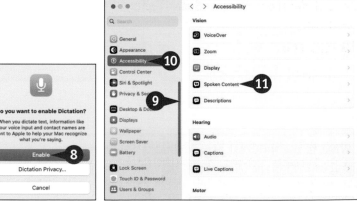

The Spoken Content pane opens.

12 Click the **System speech language** pop-up menu (⇕) and click the language to use.

13 Click the **System voice** pop-up menu (⇕) and select the voice you prefer.

14 Click **Play Sample** to hear the voice.

15 Drag the **Speaking rate** slider to adjust the speed if necessary.

16 Drag the **Speaking volume** slider to adjust the volume if necessary.

17 Select **Speak announcements** (☑) to hear alerts.

18 Select **Speak selection** (☑) to make your MacBook speak selected text.

19 Select **Speak items under the pointer** (☑) to make your MacBook speak items when you hold the pointer over them.

20 Select **Speak typing feedback** (☑) to hear the characters you type.

Note: You can click **Options** (ⓘ) to configure settings for Speak Announcements, Speak Selection, Speak Items Under the Pointer, or Speak Typing Feedback.

21 Click **Close** (⬤).

System Settings closes.

TIP

How can I add other system voices?
Click **System Settings** (⚙) on the Dock to open the System Settings window, click **Accessibility** (♿) in the sidebar to display the Accessibility pane, and then click **Spoken Content** (🗨) to display the Spoken Content pane. Click the **System Voice** pop-up menu (⇕), and then click **Manage Voices** to open the Manage Voices dialog. Click your language in the sidebar to see the available voices. You can then click Play (▶) to listen to the sample of a voice and click **Download** (⬇) to download the voice.

Using Dictation and Spoken Content

After enabling and configuring the Dictation and Spoken Content features on your MacBook, you can use them freely as you work or play. When you are using an app that accepts text input, you can press your keyboard shortcut to turn on Dictation and then dictate text into a document.

To make Spoken Content read to you, you select the text you want to hear and then press the appropriate keyboard shortcut. If you have enabled the announcing of alerts, your MacBook automatically speaks their text as well.

Using Dictation and Spoken Content

Using Dictation

1 Open the app into which you want to dictate text. For example, click **Notes** (▭) on the Dock to open Notes.

2 Open the document into which you will dictate text. For example, in Notes, click the appropriate note.

3 Position the insertion point.

4 Press the keyboard shortcut you set for starting Dictation, such as pressing ⊕ twice (not shown).

Ⓐ The Dictation icon appears.

5 Speak into your microphone (not shown).

Note: To enter a word with an initial capital letter, say "cap" followed by the word — for example, say "cap director" to enter "Director." To enter punctuation, say the appropriate word, such as "period," "comma," or "semicolon." To create a new paragraph, say "new paragraph."

Ⓑ Dictation inserts the text in the document.

6 Click the Dictation icon or press your shortcut again.

The Dictation icon disappears.

Make the Spoken Content Feature Read Text Aloud

1 In an app that contains text, select the text you want to hear.

2 Press the keyboard shortcut you set for Spoken Content (not shown).

Note: The default keyboard shortcut is `Option` + `Esc`.

Spoken Content reads the selected text to you.

C The highlight shows the current word.

D The control bar enables you to control playback.

Hear Alerts from Spoken Content

E After an alert appears, Spoken Content waits as long as specified in Spoken Content settings.

If you have not dismissed the alert, Spoken Content announces the app's name — for example, "Alert from Calendar" — and then reads the text of the alert.

F You can click **Close** (×) to dismiss the alert.

G You can move the pointer over the alert to display the Options button, click **Options**, and then click a Snooze command, such as **Snooze for 15 minutes**.

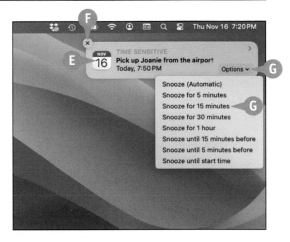

How accurate is Dictation?

Dictation can be highly accurate, but your results depend on how clearly you speak and how faithfully your microphone captures the sound. For best results, use a headset microphone and position it to the side of your mouth, outside your breath stream, to avoid distortion.

When reviewing dictated text, read it for sense to identify incorrect words and phrases. Because all the text is spelled correctly, it can be easy to overlook mistakes caused by Dictation inserting the wrong words or phrases instead of what you said.

Configure and Invoke Siri

The powerful Siri feature on your MacBook enables you to take essential actions by using your voice to tell your MacBook what you want. Siri requires an Internet connection, because the speech recognition runs on servers in Apple's data centers. You can configure several settings to specify how to invoke Siri and to make Siri work your way. Apple anonymizes the data it captures from your use of Siri and Dictation, so your privacy is not compromised; you can also delete your Siri and Dictation history, as explained in the second tip in this section.

Configure and Invoke Siri

Configure Siri

1 Click **System Settings** (⚙) on the Dock.

The System Settings window opens.

2 Click **Siri & Spotlight** (▣) in the sidebar.

The Siri & Spotlight pane appears.

3 Set the **Ask Siri** switch to On (⬤).

4 Click the **Listen for** pop-up menu (⬍), and then click the **"Siri" or "Hey Siri"** item or the **"Hey Siri"** item. If you do not want to use Siri, click **Off**.

5 Set the **Allow Siri when locked** switch to On (⬤) if you want to use Siri from the lock screen.

6 Click the **Keyboard shortcut** pop-up menu (⬍) and then click the shortcut you want to use.

7 Click the **Language** pop-up menu (⬍) and then click your language, such as **English (United States)**.

8 Click **Select** to open the dialog for selecting a voice.

9 In the Voice Variety box, click the variety, such as **American**.

10 In the **Siri Voice** box, click the voice to play a sample.

11 Click **Done** to close the dialog.

12 Click **Siri Suggestions & Privacy**.

126

The Siri Suggestions & Privacy dialog opens.

13 In the sidebar, click the app or feature you want to configure.

The settings for that app or feature appear.

14 Set the **Show Siri Suggestions in application** switch to On (⬤) or Off (⬤), as needed.

15 Set the **Learn from this application** switch to On (⬤) or Off (⬤), as needed.

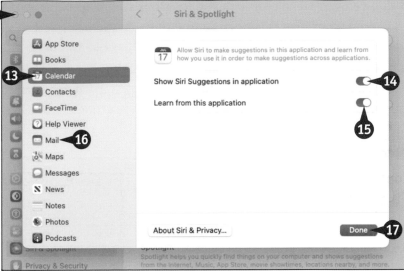

16 Repeat steps **13** to **15** for each app.

17 Click **Done** to close the Siri Suggestions & Privacy dialog.

18 Click **Close** (⬤) to close System Settings.

Invoke Siri

1 Say "Hey Siri" or "Siri" or click **Siri** (⬤) on the menu bar.

Ⓐ The Siri icon appears, and a tone plays to tell you that Siri is ready to help.

Note: If you invoke Siri unintentionally, press Esc to close the Siri window.

2 Tell Siri what you want to do or research (not shown). See the next section for some of your options.

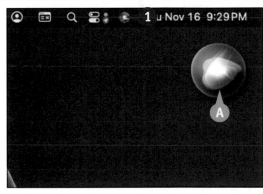

How do I put Siri on the menu bar?
Click **System Settings** (⬤) to open the System Settings window, and then click **Control Center** (⬤) to display the Control Center pane. In the Menu Bar Only section, click the **Siri** pop-up menu (⬍), and then click **Show in Menu Bar**.

What does the Delete Siri & Dictation History button in Siri settings do?
This button enables you to delete the data on your use of Siri and Dictation from Apple's servers. Click this button to open the Delete Siri & Dictation History dialog, and then click **Delete** to confirm the deletion.

Perform Tasks with Siri

You can use Siri either with the MacBook's built-in microphone or with a headset microphone. You can get good results from the built-in microphone if you are in a quiet environment and you speak loudly and clearly, or if you speak as close to the built-in microphone as possible. Otherwise, use a headset microphone, which will usually provide better results.

This section gives examples of the many ways you can put Siri to use.

Send an E-Mail Message

Say "E-mail" and the contact's name, followed by the message. Siri creates an e-mail message to the contact and enters the text. Review the message, and then click **Send** to send it.

You can also say "E-mail" and the contact's name and have Siri prompt you for the various parts of the message. This approach gives you more time to collect your thoughts.

If your Contacts list has multiple e-mail addresses for the contact, Siri prompts you to choose the address to use.

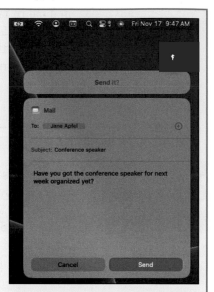

Send a Text Message

Say "Tell" or "Text" and the contact's name. When Siri responds, say the message you want to send. For example, say "Tell Kelly Wilson" and then "I'm stuck in traffic, but I should be there in about 15 minutes. Please start the meeting without me." Siri creates a text message to the contact, enters the text, and sends the message when you say "Send" or click **Send**.

You can also say "Tell" and the contact's name followed immediately by the message. For example, "Tell Bill Sykes the package will arrive at 10 AM."

Set a Reminder for Yourself

Say "Remind me" and the details of what you want Siri to remind you of. For example, say "Remind me to take the wireless transmitter to work tomorrow morning at 8 o'clock." Siri listens to what you say and creates a reminder. Check what Siri has written; if the reminder is correct, you need do nothing more, but if it is incorrect, say "Change" or click **Change** to alter it.

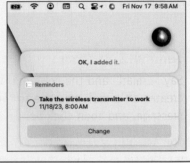

Set Up an Event

Say "Meet with" and the contact's name, followed by brief details of the appointment. For example, say "Set up a power breakfast with Don Williamson at the Crook Hotel at 8 AM on Monday." If there is a scheduling conflict, Siri warns you about it. Siri then schedules the appointment for you, adding it to your calendar.

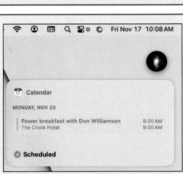

Gather Information

You can use Siri to research a wide variety of information online — everything from sports and movies to restaurants worth visiting (or worth avoiding). Here are three examples:

- "Siri, when's the next Patriots game?"
- "Siri, where is the movie *Oppenheimer* playing in Indianapolis?"
- "Siri, where's the best Mexican food in Juneau?"

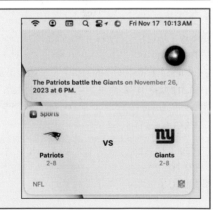

Remove Apps

m acOS enables you to remove apps you have added, but not apps included with the operating system. If you no longer need apps you have installed, you can remove them to reclaim the disk space they occupy and to prevent them from causing your MacBook to run more slowly.

You can remove most apps by simply moving them to the Trash, either from Launchpad or from the Applications folder. But if an app has an uninstall utility, you should run that utility to remove the app and its ancillary files.

Remove Apps

Remove an App Using Launchpad

1 Click **Launchpad** ( ) on the Dock.

Note: You can also place four fingers on the trackpad and pinch them together to display the Launchpad screen.

The Launchpad screen appears.

A If necessary, click a dot to move to another screen in Launchpad. You can also swipe left or right with two fingers on the trackpad.

2 Click the app and drag it to the Trash.

A confirmation dialog opens.

3 Click **Delete**.

Finder places the app in the Trash.

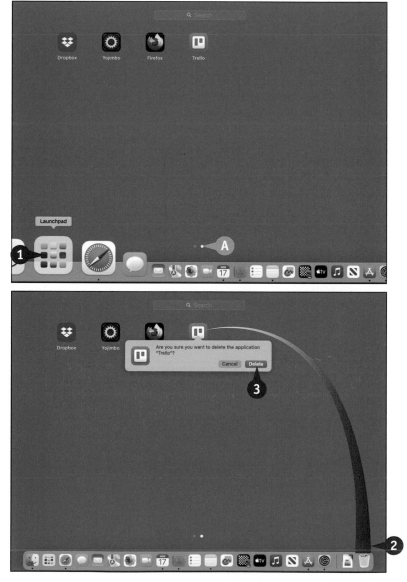

Remove an App Using the Applications Folder

1 Click **Finder** (🙂) on the Dock.

A Finder window opens.

2 Click **Applications** (A) in the sidebar or press ⌘+Shift+A.

The contents of the Applications folder appear.

3 Press Control+click the app you want to remove.

The contextual menu opens.

4 Click **Move to Trash**.

macOS moves the app to the Trash.

Where do I find the uninstall utility for an app?

If the app has a folder within the Applications folder, look inside that folder for an uninstall utility. It may also be in the Utilities folder. If there is no folder, open the disk image file, CD, or DVD from which you installed the app and look for an uninstall utility there. Some apps use an installer for both installing the app and uninstalling it, so if you do not find an uninstall utility, try running the installer and see if it contains an uninstall option. For some apps, you may need to download an uninstall utility from the app developer's website.

Identify Problem Apps

Sometimes you may find that your MacBook starts to respond slowly to your commands, even though no app has stopped working. When this occurs, you can use the Activity Monitor utility to see what app is consuming more of the processors' cycles than it should. To resolve the problem, you can quit that app and then restart it.

If you cannot quit the app normally, you can force quit it. You can force quit it either from Activity Monitor or by using the Force Quit Applications dialog.

Identify Problem Apps

1 Click **Launchpad** ( ) on the Dock.

The Launchpad screen appears.

2 Type **ac**.

Launchpad displays only those items whose names include words starting with *ac*.

3 Click **Activity Monitor** ( ).

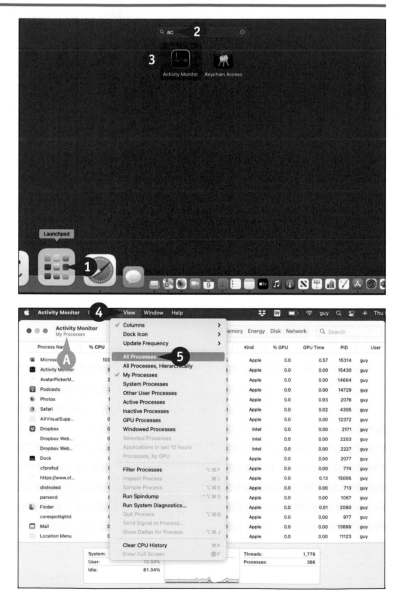

The Activity Monitor window opens, listing all your running apps and processes.

A The title bar shows (My Processes) to indicate you are viewing only your processes.

4 Click **View**.

The View menu opens.

5 Click **All Processes**.

Note: You should view all processes because another user's processes may be slowing your MacBook.

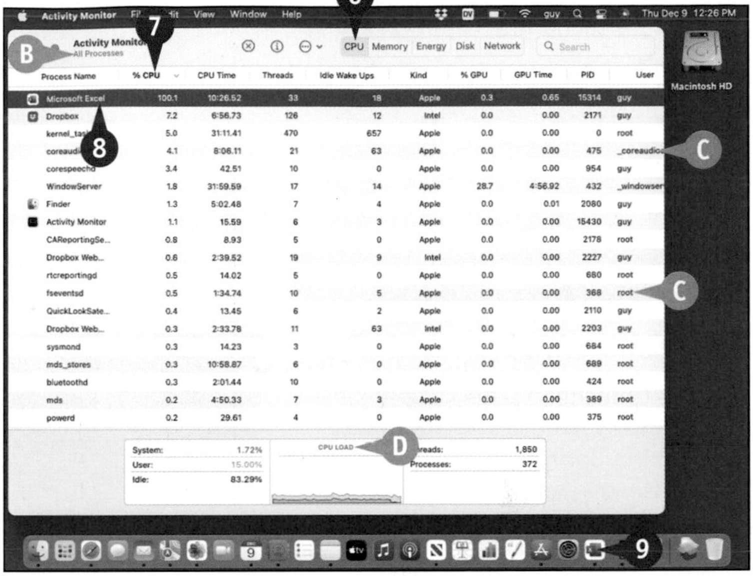

B The title bar shows (All Processes).

C Other users' processes and system processes appear as well.

6 Click **CPU**.

The details of your MacBook's central processing units, or CPUs, appear.

D The CPU Load graph shows how hard the CPU is working.

7 Click **% CPU** once or twice, as needed, so that Descending Sort (⌄) appears on the column heading.

Activity Monitor sorts the processes by CPU activity in descending order.

8 Identify the app that is using the most processor cycles.

9 Click that app's Dock icon.

The app appears.

10 Save your work in the app, and then quit it (not shown).

11 Click the Activity Monitor window.

12 Click **Activity Monitor**.

The Activity Monitor menu opens.

13 Click **Quit Activity Monitor**.

Activity Monitor closes.

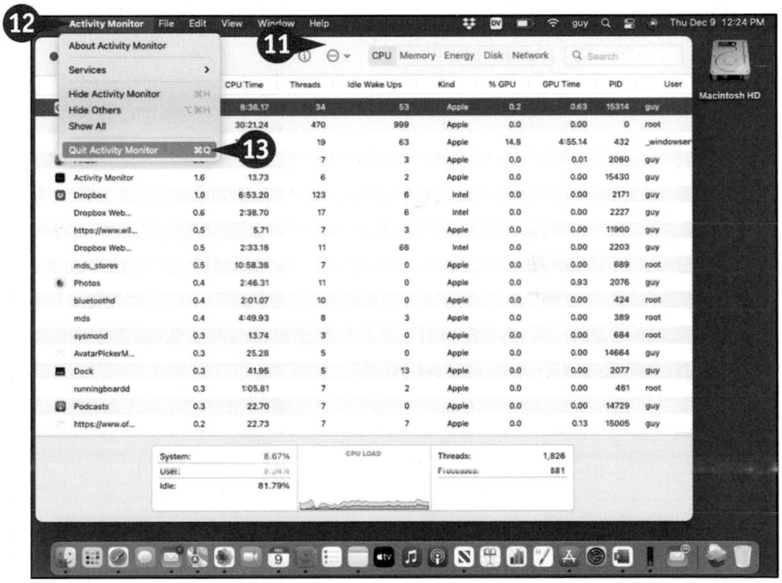

TIP

How do I see whether my MacBook is running short of memory?
Click **Memory** on the Activity Monitor tab bar and then look at the Memory Pressure readout. Click the **Memory** column heading once or twice, so that Descending Sort (⌄) appears on the column heading, to sort the processes by the amount of memory they are using.

Force a Crashed App to Quit

When an app is working normally, you can quit it by clicking the Quit command on the app's menu or by pressing ⌘+Q. But if an app stops responding to the trackpad and keyboard, you cannot quit it this way. Instead, you need to use the Force Quit command that macOS provides for this situation.

When an app stops responding, it may freeze, so that the window does not change, or it may display the spinning cursor for a long time, indicating that the app is busy.

Force a Crashed App to Quit

Force Quit an App from the Dock

1 Pressing and holding **Option**, click the app's icon on the Dock. Keep holding down the trackpad button until the pop-up menu appears.

2 Click **Force Quit**.

macOS forces the app to quit.

Force Quit an App from the Force Quit Applications Dialog

1 Click **Apple** ().

The Apple menu opens.

2 Click **Force Quit**.

Note: You can open the Force Quit Applications dialog from the keyboard by pressing **Option**+⌘+**Esc**.

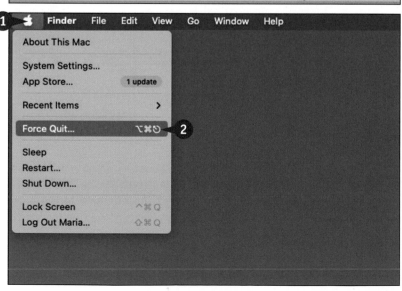

The Force Quit Applications dialog opens.

③ Click the app you want to force quit.

④ Click **Force Quit**.

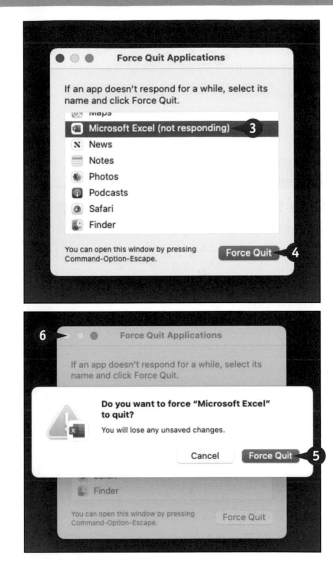

A dialog opens to confirm that you want to force quit the app.

⑤ Click **Force Quit**.

macOS forces the app to quit.

⑥ Click **Close** (●).

The Force Quit Applications dialog closes.

TIP

How do I recover the unsaved changes in a document after force quitting the app?
When you force quit an app, you normally lose all unsaved changes in the documents you were using in the app. However, some apps automatically store unsaved changes in special files called *recovery files*, which the apps open when you relaunch them after force quitting. For some apps, you may also be able to return to an earlier version of the document, covered in the next section.

Revert to an Earlier Version of a Document

macOS includes a feature called *versions* that enables apps to save different versions of the same document in the same file. You can display the different versions of the document at the same time and go back to an earlier version if necessary.

Only apps written to work specifically with the macOS versions feature enable you to revert to an older version of a document in this way. Such apps include TextEdit — the text editor and word processor included with macOS — and the apps in Apple's iWork suite.

Revert to an Earlier Version of a Document

1 In the appropriate app, open the document. For example, open a word processing document in TextEdit.

2 Click **File** on the menu bar.

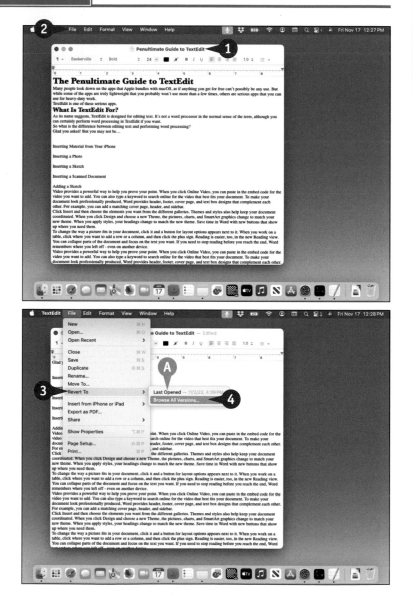

The File menu opens.

3 Click **Revert To**.

Note: You can also highlight Revert To without clicking.

The Revert To continuation menu opens.

Ⓐ You can click a version on the menu to go straight to that version.

4 Click **Browse All Versions**.

B The current version appears on the left.

C macOS displays earlier versions of the document on the right, with newer versions at the front of the stack, and older versions at the back.

5 Position the pointer over the time bars, and then click the version you want.

The version comes to the front.

6 Click **Restore**.

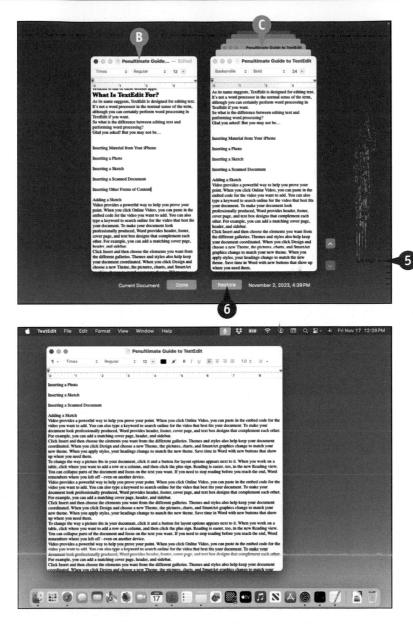

macOS restores the version of the document.

The version opens in the app so that you can work with it.

How can I tell whether an app supports versions?

With the app open, click **File** and see if the Revert To command appears on the menu. If the command appears, the app supports versions; if not, it does not.

How can I examine a document more closely before reverting to it?

When macOS displays the document versions, click the document you want to examine. The document's window enlarges so that you can see the document better. Click the background when you want to return the document window to its previous size.

CHAPTER 5

Managing Your Files and Folders

The Finder enables you to manage your files, folders, drives, and even your iPhone or iPad. You can take many actions in the Finder, including copying, moving, and deleting files and folders. You can customize the Finder settings to streamline your work.

Explore Your MacBook's File System

In macOS, your local home base is your Home folder, which is stored on your MacBook and contains folders such as Downloads, Music, and Pictures. Your online home base is your iCloud account, which macOS recommends you use to sync your Desktop folder and your Documents folder across your Macs, your iPhone, and your iPad. If you choose not to use iCloud, the Desktop folder and Documents folder appear in the Home folder on your MacBook. You can easily navigate among your folders by using the sidebar or the Go menu.

Explore Your MacBook's File System

1 Click **Finder** (　) on the Dock.

A Finder window opens to your default folder or view.

2 Click **Recents**.

The Finder window displays the Recents view, which shows the files you have used most recently.

A The Finder toolbar can display its controls as icons with text, as shown here; as icons only; or as text only. See the section "Customize the Finder Toolbar," later in this chapter, for details.

3 Click **Documents**.

Note: Documents appears in the iCloud list if you accept the macOS suggestion to store your documents in iCloud, which syncs your documents automatically across devices. If you decline this suggestion, Documents is located on your MacBook and appears in the Favorites list.

The contents of the Documents folder appear.

B The Desktop folder contains items on your desktop.

Note: The Documents folder is intended to be your storage place for word processing documents, spreadsheets, and similar files.

4 Click **Go**.

The Go menu opens.

5 Click **Home**.

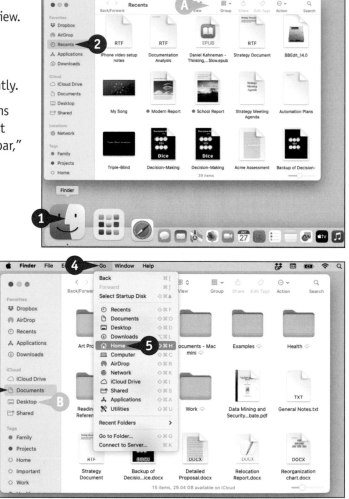

The contents of your Home folder appear.

C The Downloads folder contains files you download via apps such as Safari or Mail.

D The Dropbox folder appears only if you have installed the Dropbox online-storage app.

E The Movies folder contains movies, such as iMovie projects.

F The Pictures folder contains images.

G The Public folder is for sharing files with others.

6 Double-click the **Music** folder.

The contents of your Music folder appear.

H The Music folder contains your music library.

I The Audio Music Apps folder contains support files for GarageBand and other music apps.

J The GarageBand folder appears if you have used GarageBand, the music-composition app.

K You can click **Back** (<) to move back along the path of folders you have followed.

7 Click **Close** (●).

The Finder window closes.

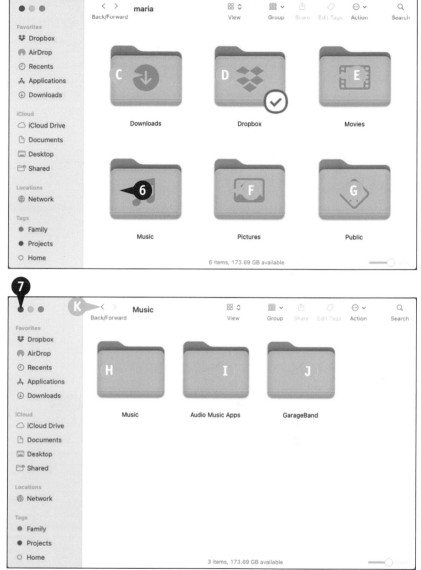

TIP

How do I choose what folder the Finder opens by default?
Click the desktop to activate the Finder. Click **Finder** to open the Finder menu, and then click **Settings**. Click **General** (⚙) to display the General pane. Click the **New Finder windows show** pop-up menu (◉), and then select the folder you want new Finder windows to display. Click **Close** (●) to close the Finder Settings window.

Using the Finder's Views

The Finder provides four views to help you find, browse, and identify your files and folders. You can switch views by clicking the View buttons on the toolbar, using the View menu or the contextual menu, or pressing keyboard shortcuts.

Icon view shows each file or folder as a graphical icon. List view shows folders as a collapsible hierarchy. Column view enables you to navigate quickly through folders and see where each item is located. Gallery view is great for identifying files visually by looking at previews of their contents.

Using the Finder's Views

Using Icon View

1 Click **Finder** (😊) on the Dock.

A Finder window opens showing your default folder or view.

2 Click the folder you want to display.

The folder appears — in this case, the Pictures folder.

3 Click **Icons** (⊞) on the toolbar.

Note: If the View pop-up menu appears instead of the four View buttons, click **View** (⊞ ◇, ≣ ◇, ▥ ◇, or ▭ ◇), and then click the appropriate view — in this case, click **as Icons** (⊞).

The files and folders appear in Icon view.

Using List View

1 Click **List** (≣) on the toolbar.

The files and folders appear in List view.

2 Click **Expand** (> changes to ∨) next to a folder.

The folder's contents appear.

Note: If the disclosure triangles do not appear next to folders, click **View** to open the View menu, and then click **Use Groups**, removing the check mark.

3 When you need to hide the folder's contents again, click **Collapse** (∨ changes to >).

Note: Click a column header in List view to sort by that column. You can click the same column header again to reverse the sort order.

Using Column View

1 In the Finder window, click **Columns** (▥) on the toolbar.

The files and folders appear in Column view.

2 Click a folder in the first column after the sidebar.

The folder's contents appear in the next column.

Note: You can click another folder if necessary.

3 Click a file.

A A preview of the file appears.

Using Gallery View

1 In the Finder window, click **Gallery** (▭) on the toolbar.

The files and folders appear in Gallery view.

2 Click a file in the thumbnail bar.

B A preview or icon appears, depending on the file type.

C Information about the selected file appears in the right pane.

Note: You can also swipe left or right on the trackpad with two fingers to move through the thumbnails on the thumbnail bar.

TIP

Can I change the size of icons used in Icon view?

Yes. The easiest way to change the size is to drag the slider on the right side of the status bar in Icon view. To display the status bar, click **View** and **Show Status Bar**.

To set a default size for Icon view, switch to Icon view and then click **View** and **Show View Options**. In the View Options window, drag the **Icon size** slider to make the icons the size you want, and then click **Use as Defaults**. Click **Close** (⬤) to close the View Options window.

Work with Finder Tabs

The Finder enables you to open multiple tabs within the same window. This capability is useful when you need to work in multiple folders at the same time. Each tab appears on the tab bar, which is below the toolbar, near the top of the window. You can navigate quickly among the tabs by clicking them.

Finder tabs are especially useful if you switch a Finder window to full-screen mode. You can drag files or folders from one Finder tab to another to copy or move the items.

Work with Finder Tabs

1 Click **Finder** (🙂) on the Dock.

A Finder window opens.

2 Click the folder you want to view in the window.

3 Press ⌘+T or click **File** and **New Tab** (not shown).

Note: The Finder hides the tab bar by default when only one tab is open. You can display the tab bar by clicking **View** and clicking **Show Tab Bar** or pressing Shift+⌘+T.

A The tab bar appears.

B A new tab opens, showing your default folder or location.

4 Click the folder you want to view.

Note: You can use a different view in each tab.

5 Click **New Tab** (+).

Note: To close a tab, position the pointer over it and then click **Close** (✕). You can also press ⌘+W or click **File** and select **Close Tab**.

144

A new tab opens.

6 Drag the tab along the tab bar to where you want it.

Note: You can drag a tab to another Finder window if you want. You can also drag a tab out of a Finder window to turn it into its own window.

7 Click **View**.

The View menu opens.

8 Click **Enter Full Screen**.

Note: Another way to switch to Full Screen view is to press Control + ⌘ + F or 🌐 + F.

The Finder window appears full screen, giving you more space for working with files, folders, and tabs.

9 Click the folder you want to display.

The folder's contents appear.

Note: To exit full-screen view, move the pointer to the top of the screen so that the menu bar appears, and then click **View** and **Exit Full Screen**. Alternatively, press Esc or press Control + ⌘ + F or 🌐 + F again.

TIP

How do I copy or move files using Finder tabs?
Select the files on the source tab, and then drag them to the destination tab on the tab bar. To put the files into the folder open in the destination tab, drop the files on the destination tab in the tab bar. To navigate to a subfolder, position the pointer over the destination tab, wait until its content appears, and then drag the items to the subfolder.

View a File with Quick Look

The Quick Look feature enables you to preview files in Finder windows without actually opening the files in their apps. You can use Quick Look to determine what a file contains or to identify the file you are looking for. You can preview a file full screen with Quick Look or preview multiple files at the same time. Quick Look works for many widely used types of files, but not for all types.

View a File with Quick Look

① Click **Finder** (🙂) on the Dock.

A Finder window opens to your default folder or view.

② Click the file you want to look through.

③ Click **Action** (⊙ ⌄).

The Action pop-up menu opens.

Note: You can also press Spacebar to open a Quick Look window for the selected item.

④ Click **Quick Look**.

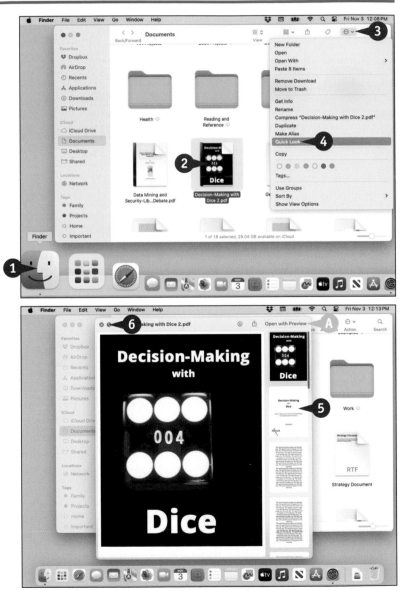

A Quick Look window opens, showing a preview of the file or the file's icon.

Note: When you use Quick Look on an audio file or a video file, macOS starts playing the file.

⑤ If you need to scroll to see more of the file, click another page thumbnail or swipe up on the trackpad.

Ⓐ You can click **Open with** to open the file in its default app.

⑥ To view the file as a full-screen preview, click **Full Screen** (◎).

The Quick Look window expands to fill the screen.

Note: To see more of the file in full-screen view, scroll down, swipe up with two fingers, or press `Page down`.

⑦ Click **Exit Full Screen** () when you finish using full-screen view.

⑧ Click **Close** (⊗) to close the Quick Look window.

Note: Instead of closing the Quick Look window, you can press →, ←, ↑, or ↓ to display another file or folder.

TIP

How do I use Quick Look on more than one file at a time?
Use Quick Look's full-screen view, which enables you to browse the files easily. Select the files you want to view with Quick Look, launch Quick Look, and click **Full Screen** (◉) to enter full-screen view. Click **Play** (▶) to play each preview for a few seconds, or click **Next** (➡) and **Previous** (⬅) to move from preview to preview. Click **Index Sheet** (▦) to see the index sheet showing all previews, and then click the item you want to see.

Organize Your Desktop Files with Stacks

M any people find the desktop a handy place to save files that they need to be able to access quickly. But saving many files to the desktop can clutter up the desktop and make files hard to find quickly.

To solve this problem, macOS provides the Stacks feature, which organizes the desktop's files by the criteria you choose. By turning on Stacks, you create a series of stacks that are easier to navigate. You can browse a stack quickly, find the file you need, and open it.

Organize Your Desktop Files with Stacks

Turn On Stacks and Choose the Grouping

1 Click anywhere on the desktop.

The Finder becomes active, and the Finder menus appear.

Note: The desktop is technically a Finder window, although it looks and behaves very differently from most Finder windows. This is why the Finder becomes active when you click the desktop.

2 Click **View**.

The View menu opens.

3 Click **Use Stacks**.

Finder sorts the documents into stacks using the default grouping, Kind. This grouping creates stacks such as Documents, PDF Documents, and Images.

4 If you want to change the grouping, click **View**.

The View menu opens.

5 Highlight **Group Stacks By** without clicking.

The Group Stacks By continuation menu opens.

6 Click the grouping you want, such as **Date Last Opened**.

Finder sorts the stacks by the new grouping.

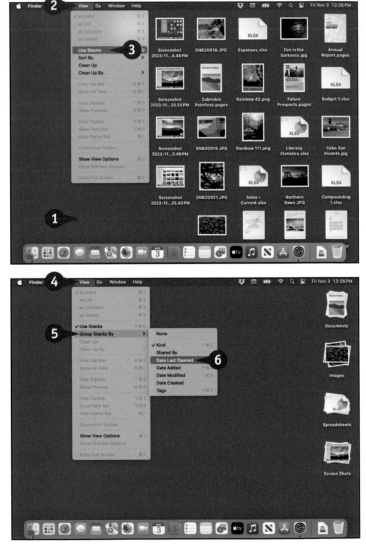

Browse a Stack and Open a File

1 Click the stack.

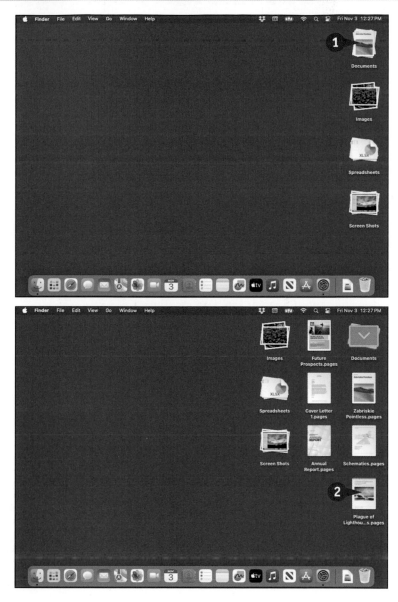

The stack's contents appear.

2 Click the file you want to open.

The file opens in its default app.

Note: You can also take other actions after displaying a stack's contents. For example, you can Ctrl +click a file and then click **Quick Look** on the contextual menu to view the file's contents using Quick Look.

What other moves can I use with Stacks?

You can quickly browse a stack by moving the pointer over the stack and dragging left or right with two fingers on the trackpad. The topmost item on the stack grows larger, so you can see its preview a little better, and its filename appears in place of the stack name. Drag left or right with two fingers to locate the item you want, and then click the item to open it.

Search for a File or Folder

macOS includes a powerful search feature called *Spotlight* that enables you to find the files and folders you need. Spotlight automatically indexes the files on your MacBook and connected drives so that it can deliver accurate results within seconds when you search.

You can use Spotlight either directly from the desktop or from within a Finder window. Depending on what you need to find, you can use either straightforward search keywords or complex search criteria.

Search for a File or Folder

Search Quickly from the Desktop

1 Click **Spotlight** (Q).

The Spotlight pop-up window opens.

Note: Spotlight indexes both the metadata and the contents of files. Metadata includes information such as the filename, file extension, and file label; the date created, date received, and date last viewed; and the subject, title, and comment assigned to the file. Contents include any text, enabling you to search by keyword in documents that contain text.

2 Type one or more keywords.

Spotlight displays a list of matches.

A The Websites section lists websites that may be relevant to your search.

B In the Related Searches section, you can click **Search the Web** (🌐) to open a Safari window that searches the Web using your search terms.

C You can click **Search in Finder** (🔍) to open a Finder window that searches your MacBook's file system using your search terms.

3 Double-click the file you want to open.

The file opens in the application associated with it.

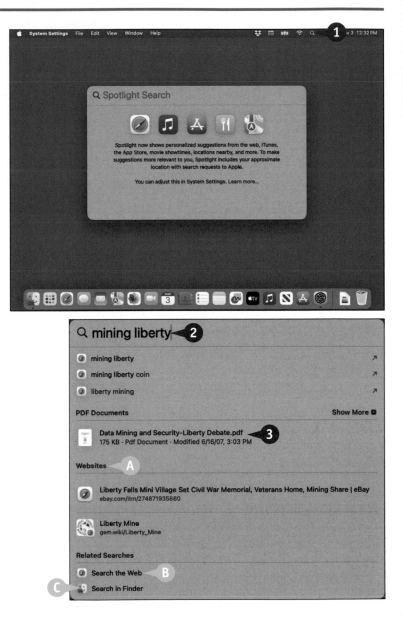

Search from a Finder Window

1 Click **Finder** ( ) on the Dock (not shown).

A Finder window opens.

2 Click **Search** (Q).

3 Type the keywords for your search.

D The Finder window's title bar changes to *Searching* and the folder or location.

E A list of search results appears.

F You can click a suggested search criterion on the pop-up menu to restrict the search.

4 To change where Spotlight is searching, click a button on the Search bar.

G You can quickly view a file by Control+clicking it and then clicking **Quick Look** or by clicking it and pressing Spacebar.

H You can open a file by double-clicking it.

I You can click **Add** (+) to add another line of search controls that enable you to refine the search.

J You can click **Save** to save the search for future use.

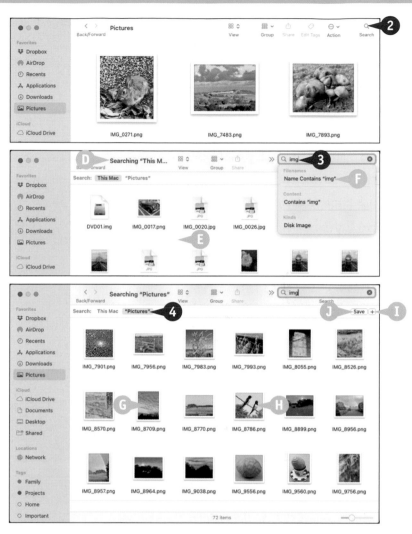

Can I change where Spotlight searches for files?

Yes. Click **System Settings** ( ) on the Dock, and then click **Siri & Spotlight** ( ) in the sidebar. At the bottom of the Siri & Spotlight pane, click **Spotlight Privacy** to display the Privacy dialog, in which you can create a list of locations you do not want Spotlight to search. To add a location, click **Add** (+), use the resulting dialog to select the location, and then click **Choose**. To remove a location, click it, and then click **Remove** (—). When the list is complete, click **Done** to close the Privacy dialog.

Create a New Folder

Y<!-- -->ou can customize the hierarchy of folders in your user account by creating as many new folders and subfolders as you need. You can create folders and subfolders in your user account or in other parts of the file system, such as on an external drive connected to your MacBook or in your iCloud account. macOS blocks you from creating folders in folders or locations for which you do not have permission.

If you want to sync your folders automatically across your devices, create the folders in your iCloud account.

Create a New Folder

1 Click **Finder** (🙂) on the Dock.

A Finder window opens to your default folder.

2 Click the folder in which you want to create the new folder.

3 Click **Action** (⊙ ⌄) on the toolbar.

The Action pop-up menu opens.

4 Click **New Folder**.

Note: You can also create a new folder by pressing ⌘+Shift+N or by clicking **File** on the Finder menu bar and clicking **New Folder**.

A A new folder appears in the Finder window.

The new folder shows an edit box around the default name, Untitled Folder.

5 Type the name you want to give the folder.

6 Press Return (not shown).

The folder takes on the new name.

7 Click or double-click the folder, depending on the view you are using.

The folder opens. You can now add files to the folder or create subfolders inside it.

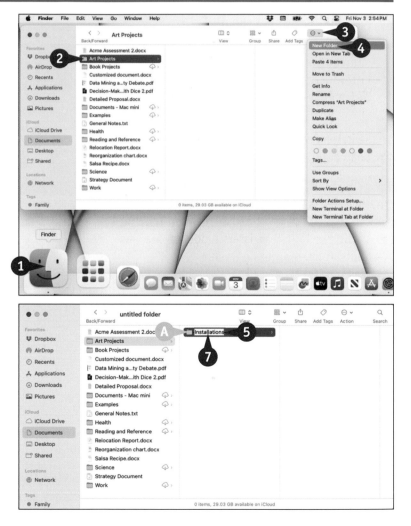

Rename a File or Folder

The Finder enables you to rename any file or folder you have created. To keep your MacBook's file system well organized, it is often helpful to rename files and folders.

macOS prevents you from renaming system folders, such as the System folder itself, the Applications folder, or the Users folder. macOS also prevents you from renaming the standard folders in each user account, such as the Documents folder and the Pictures folder, because apps expect these folders to be available with their default names.

Rename a File or Folder

1 Click **Finder** ( ) on the Dock.

A Finder window opens to your default folder.

2 Navigate to the folder that contains the file or folder you want to rename. For example, click **Documents** and then click a subfolder in the Documents folder.

3 Click the file or folder. This example uses a folder.

4 Press Return (not shown).

Note: You can also display the edit box by clicking the file's or folder's name to select it, pausing, and then clicking again. You must pause between the clicks; otherwise, the Finder registers a double-click and opens the file or folder.

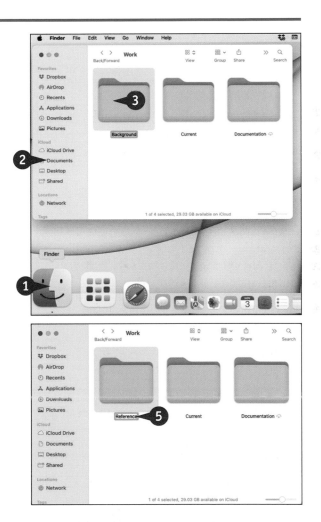

An edit box appears around the name.

5 Edit the current name, or simply type the new name over the current name.

6 Press Return (not shown).

The file or folder takes on the new name.

Note: To rename multiple files or folders at once, select them, click **File** on the menu bar, and then click **Rename**. In the Rename Finder Items dialog, choose renaming options, and then click **Rename**.

Copy a File

The Finder enables you to copy a file from one folder to another. Copying is useful when you need to share a file with other people or when you need to keep a copy of the file safe against harm.

You can copy either by clicking and dragging or by using the Copy and Paste commands. You can copy a single file or folder at a time or copy multiple items.

Copy a File

Copy a File by Clicking and Dragging

1 Click **Finder** ( ) on the Dock.

A Finder window opens.

2 Click the folder that contains the file you want to copy.

3 Click **File**.

The File menu opens.

4 Click **New Finder Window**.

A new Finder window opens.

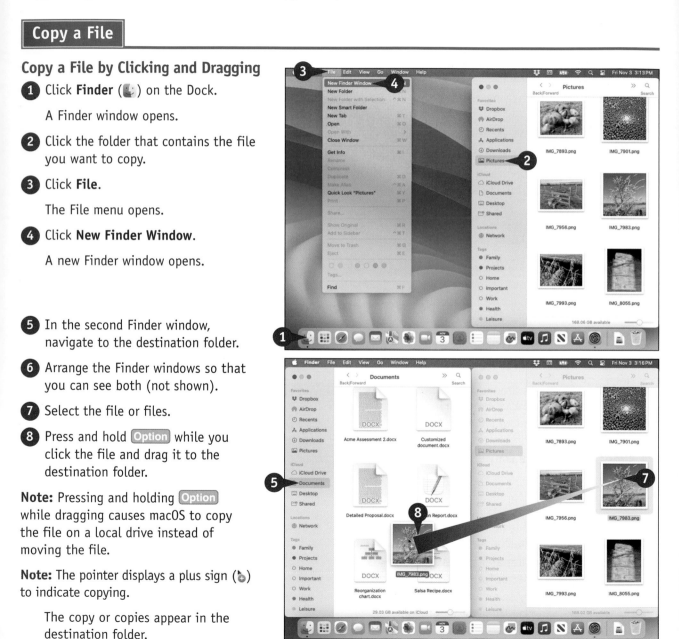

5 In the second Finder window, navigate to the destination folder.

6 Arrange the Finder windows so that you can see both (not shown).

7 Select the file or files.

8 Press and hold `Option` while you click the file and drag it to the destination folder.

Note: Pressing and holding `Option` while dragging causes macOS to copy the file on a local drive instead of moving the file.

Note: The pointer displays a plus sign ( ) to indicate copying.

The copy or copies appear in the destination folder.

Copy a File Using Copy and Paste

1 Click **Finder** () on the Dock.

A Finder window opens.

2 Click the folder that contains the file you want to copy.

3 Click the file.

Note: You can also copy the selected item by pressing ⌘+C and paste the copied or cut item by pressing ⌘+V.

4 Click **Action** (⊙ ⌄).

The Action pop-up menu opens.

5 Click **Copy**.

Finder copies the file's details to the clipboard.

6 Navigate to the folder in which you want to create the copy.

7 Click **Action** (⊙ ⌄).

The Action pop-up menu opens.

8 Click **Paste Item**.

A copy of the file appears in the destination folder.

Note: You can use the Paste command in either the same Finder window or tab or another Finder window or tab — whichever you find more convenient.

How do I copy a folder?

Use the same techniques as for files: Either press Option +drag the folder or folders to the destination folder, or use the Copy command to copy the folder and the Paste command to paste it into the destination folder.

Can I make a copy of a file in the same folder as the original?

Yes. To do this, click the file, click **Action** (⊙ ⌄), and then click **Duplicate**. Finder automatically adds *copy* to the end of the copy's filename to distinguish it from the original.

Move a File

The Finder makes it easy to move a file from one folder to another. You can move a file quickly by clicking it in its current folder and then dragging it to the destination folder.

When the destination folder is on the same drive as the source folder, the Finder moves the file to that folder. But when the destination folder is on a different drive, the Finder copies the file by default. To override this and move the file, you press ⌘ as you drag.

Move a File

Move a File Between Folders on the Same Drive

1 Click **Finder** (🙂) on the Dock.

A Finder window opens.

2 Click the folder that contains the file you want to move.

3 Click **File**.

The File menu opens.

4 Click **New Finder Window**.

A new Finder window opens.

5 In the second Finder window, navigate to the destination folder.

6 Arrange the Finder windows so that you can see both (not shown).

7 Click the file and drag it to the destination folder.

The file appears in the destination folder and disappears from the source folder.

Move a File from One Drive to Another

1 Click **Finder** (🙂) on the Dock.

A Finder window opens.

2 Click the folder that contains the file you want to move.

3 Press **Control**+click **Finder** (🙂) on the Dock.

The contextual menu opens.

A You can click a recent folder to open that folder.

4 Click **New Finder Window**.

A new Finder window opens.

5 In the second Finder window, click the drive to which you want to copy the file.

6 Click the destination folder.

7 Arrange the Finder windows so that you can see both (not shown).

8 Click the file, then press and hold ⌘ while you drag it to the destination folder.

The file appears in the destination folder and disappears from the source folder.

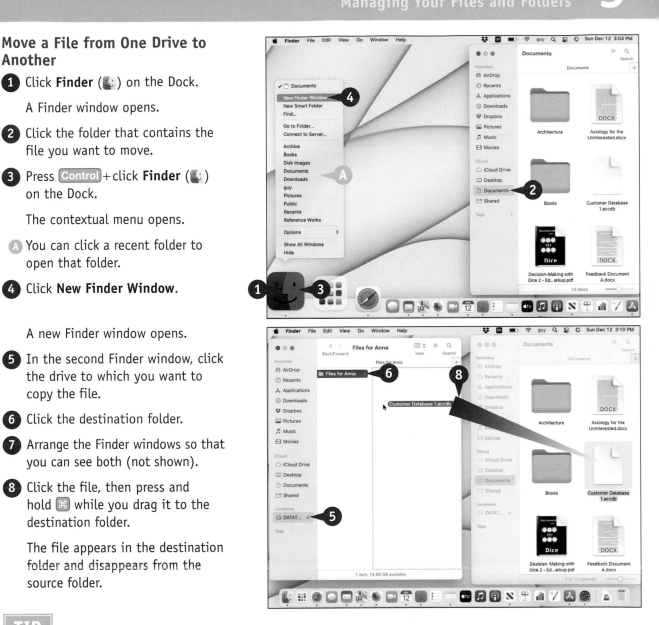

TIP

Can I move files by using menu commands rather than clicking and dragging?

Yes. If you find it awkward to drag files from one folder to another, you can use menu commands instead. Select the file or files you want to move, and then click **Edit** and **Copy** to copy them. Open the destination folder and click **Edit** to open the Edit menu. Press and hold **Option** to make Finder replace the Paste Item command or the Paste Items command with the Move Item Here command or the Move Items Here command. Then click **Move Item Here** or **Move Items Here**.

View the Information About a File or Folder

macOS keeps a large amount of information about each file and folder. When you view the file or folder in most Finder views, you can see the item's name and some basic information about it, such as its kind, size, and date last modified.

To see more information about the file or folder, you can open the Info window. This window contains multiple sections that you can expand by clicking **Expand** (>) or collapse by clicking **Collapse** (⌄).

View the Information About a File or Folder

1 Click **Finder** (🙂) on the Dock.

A Finder window opens.

2 Click the folder that contains the file whose info you want to view.

3 Click the file.

4 Click **Action** (⊙ ⌄).

The Action pop-up menu opens.

5 Click **Get Info**.

Note: You can also open the Info window for the selected item by pressing ⌘ + ⓘ.

The Info window opens.

6 View the preview.

7 Review the tags. To add a tag, type the tag's text, and then press **Return**.

Note: See the next section, "Organize Your Files with Tags," for more on tags.

8 Review the general information: *Kind* shows the file's type. *Size* shows the file's size on disk. *Where* shows the folder that contains the file. *Created* shows when the file was created. *Modified* shows when the file was last changed.

⑨ Review the details in the More Info section.

Note: The More Info details are especially useful for photos.

Ⓐ You can change the filename or extension. Normally, though, it is best not to change the extension.

⑩ Select **Hide extension** (☑) if you want to hide the extension.

⑪ Type any comments to help identify the file.

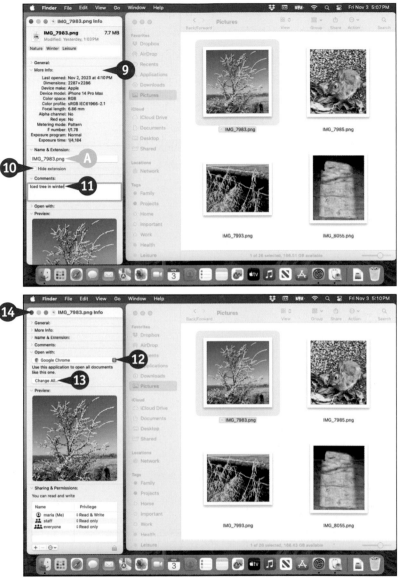

⑫ Click **Open with** (◆) and select the app with which to open this file.

⑬ Click **Change All** if you want to use the app for all files of this type.

⑭ Click **Close** (●).

The Info window closes.

TIP

How do I use the Sharing & Permissions settings?

To adjust the permissions for the file or folder, click the Privilege row for the user or group you want to affect. You can then select the appropriate level of permissions: **Read & Write**, **Read only**, or **No Access**. Click **Add** (+) to add a user or group to the list. Click **Remove** (−) to remove the selected user or group from the list.

Organize Your Files with Tags

Y ou can organize your files and folders by giving them descriptive names and storing them in appropriate locations. But macOS and its apps give you another means of organizing your files and folders: tags.

macOS includes a set of default tags that you can customize to better describe your projects. You can then apply one or more tags to a file to enable you to locate it more easily either on your MacBook or in iCloud.

Organize Your Files with Tags

Customize Your Tags

1 Click the desktop.

The Finder becomes active.

2 Click **Finder**.

The Finder menu opens.

3 Click **Settings**.

The Finder Settings window opens.

4 Click **Tags** (◇).

The Tags pane appears.

5 Click a tag you want to rename, type the new name in the edit box that appears, and then press Return.

6 Click the check box (⊟ or ☐ changes to ☑) to make the tag appear in the list in the Finder.

7 Drag the tags into the order in which you want them to appear.

8 Drag tags to the Favorite Tags list at the bottom to control which tags appear in Finder menus.

A You can click **Add** (+) to add a new tag to the list.

9 Click **Close** (●).

The Finder Settings window closes.

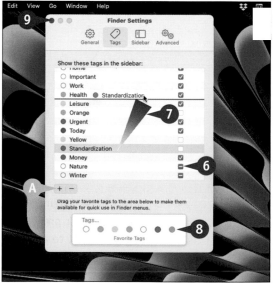

Apply Tags to Files and Folders

1 If the Tags section of the sidebar is collapsed, position the pointer over Tags and click **Expand** (**>**).

2 Click the file or folder and drag it to the appropriate tag.

Finder applies the tag.

Note: You can also apply tags from the File menu or from the contextual menu.

View Files and Folders by Tags

1 If the Tags section of the sidebar is collapsed, position the pointer over Tags and click **Expand** (**>**) when it appears.

2 Click the appropriate tag.

B The Finder window shows the tagged files and folders.

How do I apply tags to a new document I create?

In the app, click **File** and **Save** or press ⌘ + S to display the Save As dialog. Type the filename; then click **Tags** and click each tag you want to apply. You can then choose the folder in which to save the document and click **Save** to save it.

161

Work with Zip Files

macOS includes a compression tool that enables you to shrink files. Compression is especially useful for files you need to transfer across the Internet, place on a limited-capacity medium such as a USB drive, or archive for storage.

Using the Finder, you can compress a single file or multiple files. Compressing creates a compressed file in the widely used Zip format, often called a *Zip file*, that contains a copy of the files. The original files remain unchanged.

Work with Zip Files

Compress Files to a Zip File

1 Click **Finder** ( ) on the Dock.

A Finder window opens.

2 Click the folder that contains the file or files you want to compress.

3 Select the file or files.

4 Click **Action** ( ).

The Action pop-up menu opens.

5 Click **Compress**.

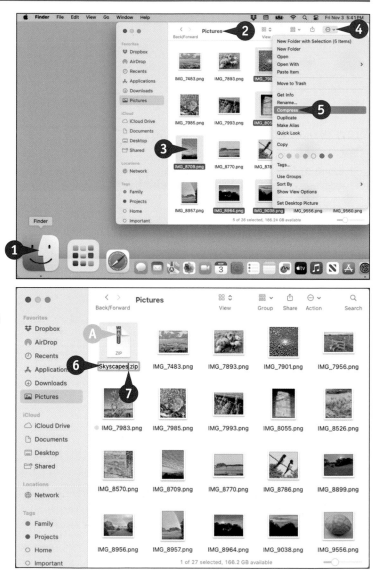

Ⓐ The compressed file appears in the folder.

Note: If you selected one file, macOS gives the file the same name with the .zip extension. If you selected multiple files, macOS names the Zip file Archive.zip.

6 Click the file and press Return.

An edit box appears.

7 Type the new name and press Return.

The file takes on the new name.

Extract Files from a Zip File

 Click **Finder** ( ) on the Dock.

A Finder window opens.

2 Navigate to the folder that contains the Zip file.

Note: If you receive the Zip file attached to an e-mail message, save the file as explained in Chapter 7.

3 Double-click the Zip file.

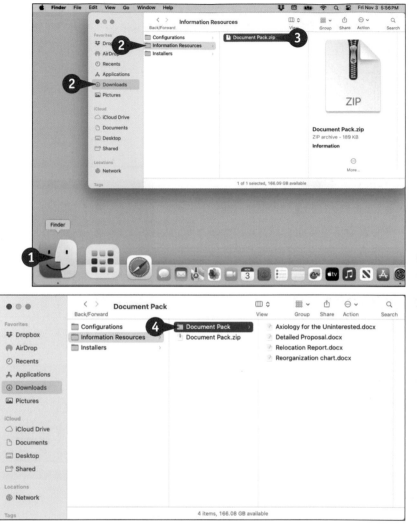

Archive Utility unzips the Zip file, creates a folder with the same name as the Zip file, and places the contents of the Zip file in it.

4 Click the new folder to see the files extracted from the Zip file.

Note: If you no longer need the Zip file once you have extracted its contents, you can delete the Zip file.

When I compress a music file, the Zip file is bigger than the original file. What have I done wrong?
You have done nothing wrong. Compression removes extra space from the file and can squeeze some graphics and text files down by as much as 90 percent. But if you try to compress an already compressed file, such as an MP3 audio file or an MPEG video file, Archive Utility cannot compress it further — and the Zip file packaging adds a small amount to the file size.

Using the Trash

macOS provides a special folder called the Trash in which you can place files and folders you intend to delete. Like a real-world trash can, the Trash retains files until you actually empty it. So if you find you have thrown away a file that you need after all, you can recover the file from the Trash. The Trash icon appears at the right end of the Dock by default, giving you quick access to the Trash and an easy way to eject removable media, which you must remove from the macOS file system before physically disconnecting.

Using the Trash

Place a File in the Trash

1 Click **Finder** (🙂) on the Dock.

A Finder window opens to your default folder or view.

2 Click the folder that contains the file you want to throw in the Trash.

3 Click the file you want to delete.

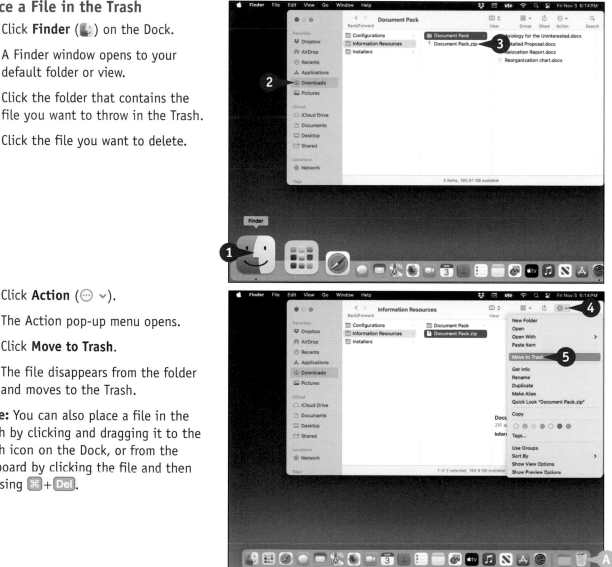

4 Click **Action** (⊙ ⌄).

The Action pop-up menu opens.

5 Click **Move to Trash**.

Ⓐ The file disappears from the folder and moves to the Trash.

Note: You can also place a file in the Trash by clicking and dragging it to the Trash icon on the Dock, or from the keyboard by clicking the file and then pressing ⌘+Del.

Recover a File from the Trash

1 Click **Trash** (🗑) on the Dock.

The Trash window opens.

B You can click **Empty** to empty the Trash. In the confirmation dialog that opens, click **Empty Trash**.

Note: You can use Quick Look to examine a file in the Trash. For example, click the file and press `Spacebar` to open the Quick Look window. However, you cannot open a file from the Trash; to open a file, you must first remove it from the Trash, putting it in either its original folder or another folder.

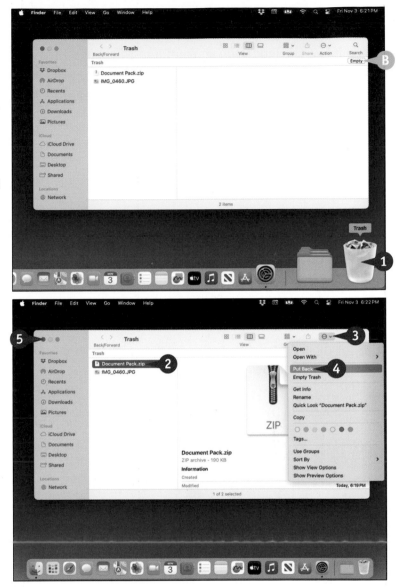

2 Click the file you want to recover.

3 Click **Action** (⊙ ∨).

The Action pop-up menu opens.

4 Click **Put Back**.

The Finder restores the file to its previous folder.

Note: If you want to put the file in a different folder, drag it to that folder. For example, drag the file to the desktop.

5 Click **Close** (●) or press ⌘+W.

The Trash window closes.

TIP

What else can I do with the Trash?

You can use the Trash to eject a removable disk, CD, or DVD in an optical drive you have connected to your MacBook. When you click a removable disk, CD, or DVD and drag it toward the Trash, the Trash icon changes to an Eject icon (⏏). Drop the item on the Eject icon to eject it. When you drag a recordable CD or DVD to which you have added files toward the Trash, macOS displays a Burn icon (☢). Drop the disc on the Burn icon to start burning it.

Configure Finder Settings

The Finder is the application that controls the macOS desktop, how files and folders are managed, and many other aspects of the way your MacBook operates. Like most applications, the Finder has a suite of settings you can configure to change the way it looks and works. You change Finder settings using its Settings command. The Settings window has several tabs that you use to configure specific aspects of how the Finder looks and behaves.

Configure Finder Settings

1. Click the desktop to make the Finder active.

2. Click **Finder** to open the Finder menu.

3. Click **Settings** to open the Finder Settings window.

4. Click **General** (⚙️) to display the General tab.

5. In the Show These Items on the Desktop list, select (☑️) for each item you want to appear on the desktop.

6. Click **New Finder windows show** (↕️) and select the default location for new Finder windows, such as **Recents**.

7. Select **Open folders in tabs instead of new windows** (☑️) if you want each folder to open in a new tab in the current window rather than in a new window.

8. Click **Sidebar** (▭).

The Sidebar tab appears.

Note: The Sidebar tab in Finder settings displays the default list of items for the sidebar. To add or remove individual locations or files, use the technique explained in the section "Customize the Finder Siderbar," later in this chapter.

9. Select (☑️) each item you want the sidebar to show.

10. Deselect (☐) each item you want to remove from the sidebar.

11. Click **Advanced** (⚙️).

The Advanced tab appears.

12 Select (☑) **Show all filename extensions** to make the Finder always display all filename extensions.

13 Select (☑) **Show warning before changing an extension** to receive a warning when you change a filename extension. This is normally helpful.

14 Select (☑) **Show warning before removing from iCloud Drive** to receive a warning when you move a file so that it will no longer be synced by iCloud Drive.

15 Select (☑) **Show warning before emptying the Trash** to confirm emptying the Trash.

16 Select (☑) **Remove items from the Trash after 30 days** if you want macOS to delete items automatically after 30 days.

17 In the Keep Folders on Top area, select (☑) each check box if you want folders to remain at the top of windows when sorting items by name or on the Desktop.

18 Click **When performing a search** (⬍) and then click **Search This Mac**, **Search the Current Folder**, or **Use the Previous Search Scope**, as needed.

19 Click **Close** (●).

Which folder should I use as my start folder for new Finder windows?

Choose whichever folder you find most convenient — for example, the folder in which you keep your most important documents. Click **New Finder windows show** (⬍) and select **Other.** In the dialog that opens, click the folder you want to use, and then click **Choose.**

Customize the Finder Toolbar

The toolbar that appears at the top of the Finder window contains buttons that you can use to access commands quickly and easily. The Finder toolbar includes a useful set of buttons by default, but you can configure the toolbar so that it contains the buttons you use most frequently. The Path button enables you to easily see and navigate the folder to the path the Finder window shows. The Space item and the Flexible Space item enable you to add spacing to the controls on the toolbar.

Customize the Finder Toolbar

1. Click **Finder** (😀) on the Dock.

   A Finder window opens to your default folder or view.

2. Click **View** to open the View menu.

3. Click **Customize Toolbar** to open the Customize Toolbar dialog.

4. To add a button to the toolbar, drag it from the Customize Toolbar dialog to the toolbar.

Note: You can click a button on the toolbar and drag it to a new position.

Ⓐ To restore the toolbar to its original state, drag the **. . . or drag the default set into the toolbar** box to the toolbar.

5. To remove a button from the toolbar, drag its icon from the toolbar into the Customize Toolbar dialog.

6. Click **Show** (◇) and then click the appearance you want: **Icon and Text**, **Icon Only**, or **Text Only**.

7. Click **Done**.

   The Customize Toolbar dialog closes.

   The Finder toolbar appears in its customized form.

Customize the Finder Sidebar

The sidebar on the left side of a Finder window gives you quick access to files, folders, and apps. You can customize the sidebar from its default contents to make it contain only the items you find most useful.

The sidebar contains four sections. Favorites are items you access frequently. iCloud includes your iCloud Drive, plus your Desktop folder and Documents folder if you choose to store these in iCloud. Locations include your MacBook, hard disks, external disks and optical drives, computers running the Bonjour sharing protocol, and connected servers. Tags shows your list of tags for identifying and accessing items.

Customize the Finder Sidebar

① Click **Finder** (🙂) on the Dock.

A Finder window opens to your default folder or view.

Note: If the sidebar does not appear, click **View** and then click **Show Sidebar**. Alternatively, press ⌘+Option+S.

Ⓐ You can move the pointer over the border of the sidebar, and then drag to change the width.

② Click the folder that contains the item you want to add to the sidebar.

③ Click the item you want to add.

④ Click **File**.

The File menu opens.

⑤ Click **Add to Sidebar**.

The item appears at the bottom of the sidebar.

You can then drag the new item up the sidebar to a different position.

Note: To remove an item from the sidebar, press Control+click the item, and then click **Remove from Sidebar** on the contextual menu.

CHAPTER 6

Surfing the Web

If your MacBook is connected to the Internet, you can browse, or *surf*, the sites on the World Wide Web. For surfing, macOS provides a web browser app called Safari. Using Safari, you can quickly move from one web page to another, search for interesting sites, and download files to your MacBook.

Open Safari and Customize Your Start Page

afari is the default browser for macOS. When you launch Safari, it displays the start page, a customizable page that can contain a background image and up to eight categories of links and information: Favorites, which you designate; Frequently Visited, which Safari derives automatically as you browse; Shared with You, links from other people; Privacy Report, details of threats to your privacy; Siri Suggestions, links from Apple's virtual assistant; Reading List, a list of pages you mark for reading; Recently Closed Tabs, tabs you have closed recently; and iCloud Tabs, pages from your other devices that sign in to iCloud.

Open Safari and Customize Your Start Page

1 Click **Safari** (🧭) on the Dock.

Safari opens and shows your start page.

Ⓐ The Favorites section displays favorites you have created for web pages.

Ⓑ The Frequently Visited section displays sites you have visited frequently.

Note: When you start using Safari, the Frequently Visited section will be empty, and Safari may hide it automatically.

Ⓒ The Shared with You section shows links that others have shared with you.

Ⓓ Further down the start page, the Privacy Report section shows how many trackers Safari has blocked from following you in the last 7 days.

Ⓔ The Siri Suggestions section shows website suggestions from Siri.

Ⓕ The Reading List section shows web pages you have added to your Reading List.

Ⓖ The iCloud Tabs section shows web pages open on your other devices.

Ⓗ You can click ◇ and select the iCloud device.

Ⓘ You can click **Show all** (⌄) to display all tabs for the device. Click **Show less** (⌃) to show fewer tabs.

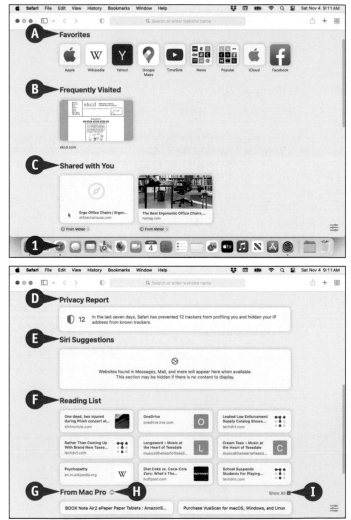

2 Click **Settings** (☰).

The Settings pop-up menu opens.

3 Select (☑) or deselect (☐) **Use Start Page on All Devices**, as needed.

4 Select (☑) or deselect (☐) **Favorites** (☆), **Frequently Visited** (🕐), **Shared with You** (👥), **Privacy Report** (🛡), **Siri Suggestions** (⊗), **Reading List** (∞), **Recently Closed Tabs** (🗐), and **iCloud Tabs** (☁), as needed.

5 To apply a background image to the start page, select **Background Image** (🖼).

The Background Image section of the pop-up menu opens.

6 In the scrolling list, click the background image you want to apply.

J Safari applies the image.

7 Click outside the pop-up menu.

The pop-up menu closes.

TIP

How do I use an image of my own for the background of the start page?
Select **Background Image** (☑) to display the Background Image section of the Settings pop-up menu, and then click **+** at the beginning of the list of background image thumbnails. In the dialog that opens, navigate to the appropriate folder, such as your Pictures folder, and then click **Choose**.

Open a Web Page

Safari enables you to browse the Web in various ways. The most straightforward way to reach a web page is to type or paste its unique address, which is called a *uniform resource locator* or *URL*, into the address box in Safari.

This technique works well for short addresses but is slow and awkward for complex addresses. Instead, you can click a link or click a bookmark for a page you have marked. If you do not have a bookmark, you can use a search engine, such as Google or Bing, to locate the address.

Open a Web Page

1 Click **Safari** (🧭) on the Dock.

Safari opens.

2 Click anywhere in the address box or press ⌘+L.

Safari selects the current address.

3 Type the URL of the web page you want to visit.

Note: You do not need to type the http:// or https:// part of the address. Safari adds this automatically for you when you press Return.

4 Press Return (not shown).

Safari displays the web page.

Note: To search, click the address box and start typing your search terms. If the Suggestions list displays a result that is what you want, click that result to go to the page. Otherwise, finish typing your search terms, press Return to display a page of search results, and then click the result you want to see.

A If the website displays a notice about cookies or data storage, use the controls to choose which cookies to accept.

Note: *Cookies* are small text files websites use to store data about users and to track their activity.

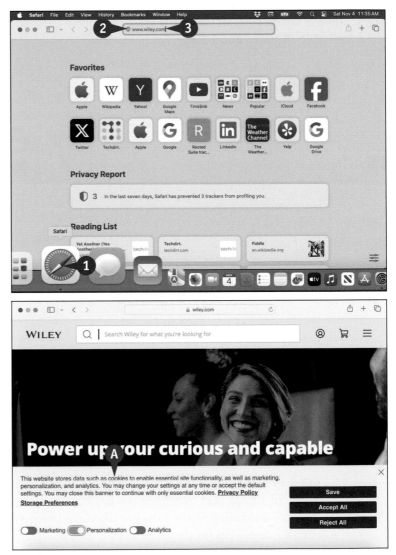

Follow a Link to a Web Page

You can click a link on a web page in Safari to navigate to another page or another marked location on the same page. Most web pages contain multiple links to other pages, which may be either on the same website or on another website. Some links are underlined, whereas others are attached to graphics or to different-colored text. When you position the pointer over a link, the pointer changes from the standard arrow (▶) to a hand with a pointing finger (☝).

Follow a Link to a Web Page

1 In Safari, position the pointer over a link (▶ changes to ☝).

Note: If you want to see the address of a link over which you move the pointer, display the status bar by clicking **View** and then clicking **Show Status Bar**. The address appears in the status bar.

2 Click the link.

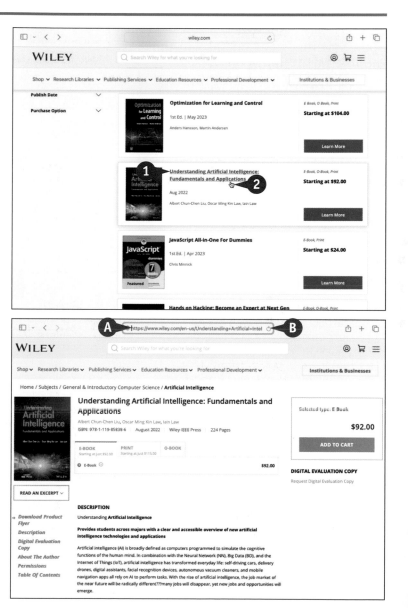

Safari shows the linked web page.

Ⓐ You can click the address box to see the full address of the web page to which you have navigated.

Note: The address box in Safari normally shows the base name of the website you are visiting, such as wiley.com, rather than the full address of the web page.

Ⓑ You can click **Reload** (↻) to reload the web page. You might reload a web page if it does not fully load or if you think new information may now be available.

Open Several Web Pages at Once

Safari enables you to open multiple web pages at the same time, which is useful for browsing quickly and easily. You can open multiple pages either on separate tabs in the same window or in separate windows. You can drag a tab from one window to another.

Use separate tabs when you need to see only one of the pages at a time. Use separate windows when you need to compare two pages side by side.

Open Several Web Pages at Once

Open Several Pages on Tabs in the Same Safari Window

1 Go to the first page you want to view.

Note: You can also click **Add** (**+**) in the upper-right corner of the Safari window or press ⌘+T to open a new tab showing your home page. Type a URL in the address box, and then press Return to go to the page.

2 Press Control+click a link.

The contextual menu opens.

3 Click **Open Link in New Tab**.

A Safari opens the linked web page in a new tab.

Note: You can repeat steps **2** and **3** to open additional pages on separate tabs.

4 To change the page Safari is displaying, click the tab for the page you want to see.

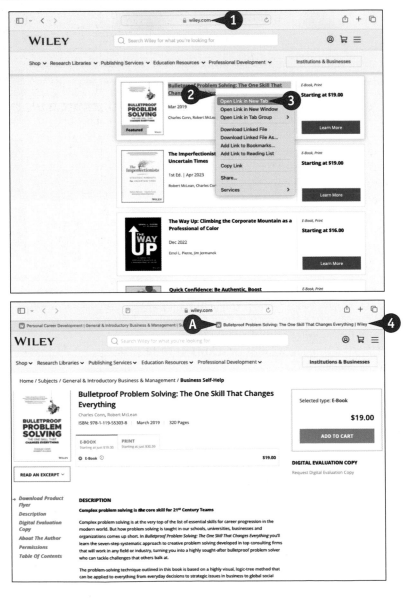

Open Several Pages in Separate Safari Windows

1 Go to the first page you want to view.

2 Press **Control** + click a link.

The contextual menu opens.

3 Click **Open Link in New Window**.

Note: You can also open a new window by pressing ⌘ + **N**.

B Safari opens the linked web page in a new window.

4 To move back to the previous window, click it. If you cannot see the previous window, click **Window** and then click the window on the Window menu.

Note: You can also move back to the previous window by closing the new window you just opened.

Note: Press ⌘ + ˜ to cycle forward through windows and ⌘ + **Shift** + ˜ to cycle backward.

Note: You can click **Window** and **Merge All Windows** to merge all Safari windows onto tabs in a single window.

Can I change the way Safari tabs and windows behave?
Yes. Click **Safari** and **Settings** to open the Settings window, and then click **Tabs** (⬜). Select **Automatically collapse tab titles into icons** (☑) to reduce tab titles to icons when space is tight. Click **Open pages in tabs instead of windows** (⬦) and then click **Never**, **Automatically**, or **Always**, as needed. Select **⌘+click to open a link in a new tab** (☑) to use ⌘ for opening a new tab. Select **When a new tab or window opens, make it active** (☑) if you want to switch to the new tab or window on opening it. Select **Use ⌘+1 through ⌘+9 to switch tabs** (☑) to switch tabs quickly by pressing shortcuts. Click **Close** (⬤) to close the Settings window.

Navigate Among Web Pages

Safari makes it easy to navigate among the web pages you browse. Safari tracks the pages that you visit so that the pages form a path. You can go back along this path to return to a page you viewed earlier; after going back, you can go forward again as needed.

Safari keeps a separate path of pages in each open tab and each open window so you can move separately in each tab or window. You can also navigate using the browser history, as explained in the next section, "Return to a Recently Visited Page."

Navigate Among Web Pages

Go Back One Page

1 In Safari, click **Previous Page** (**<**).

Note: You can also swipe right with two fingers on the trackpad to go back to the previous page. You must start the movement from the left side of the window.

Safari displays the previous page you visited in the current tab or window.

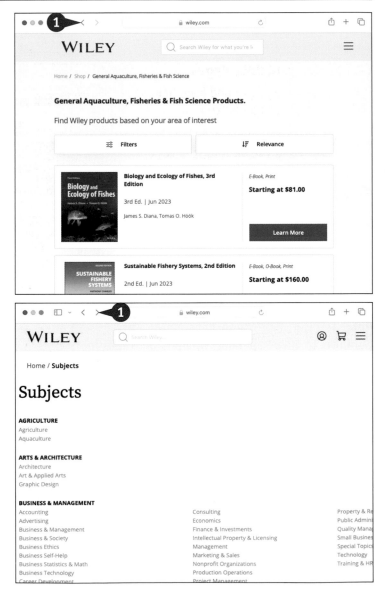

Go Forward One Page

1 Click **Next Page** (**>**).

Note: The Next Page button is available only when you have gone back. Until then, there is no page for you to go forward to.

Note: You can also swipe left with two fingers on the trackpad to go forward to the next page. You must start the movement from the right side of the window.

Safari displays the next page for the current tab or window.

Go Back Multiple Pages

1 Click **Previous Page** (<) and keep holding down the trackpad button.

A pop-up menu opens showing the pages you have visited in the current tab or window.

2 Click the page you want to visit.

Safari displays the page.

Go Forward Multiple Pages

1 Click **Next Page** (>) and keep holding down the trackpad button.

A pop-up menu opens showing the pages further along the path for the current tab or window.

2 Click the page you want to visit.

Safari displays the page.

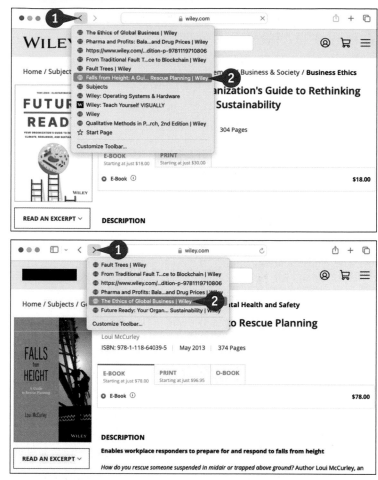

How can I navigate with the keyboard?

You can use these following keyboard shortcuts:

- Press ⌘+[to display the previous page.
- Press ⌘+] to display the next page.
- Press ⌘+Shift+H to display your home page.
- Press ⌘+Shift+[to display the previous tab.
- Press ⌘+Shift+] to display the next tab.
- Press ⌘+W to close the current tab and display the previous tab. If the window has no tabs, this command closes the window.
- Press ⌘+Shift+W to close the current window and display the previous window, if there is one.
- Press ⌘+1 to display the first tab, ⌘+2 to display the second tab, and so on up to ⌘+9.

Return to a Recently Visited Page

To help you return to web pages you have visited before, Safari keeps the History list of all the pages you have visited recently.

Normally, each person who uses your MacBook has a separate user account, so each person has their own History. But if you share a user account with other people, you can clear the History list to prevent them from seeing what web pages you have visited. You can also shorten the length of time for which History tracks your visits.

Return to a Recently Visited Page

Return to a Page on the History List

1 In Safari, click **History**.

The History menu opens.

A If a menu item for the web page you want appears on the top section of the History menu, before the time and day continuation menus, click the item.

2 Highlight or click the time or day on which you visited the web page.

The continuation menu opens, showing the sites you visited at that time or on that day.

3 Click the web page to which you want to return.

Safari displays the web page.

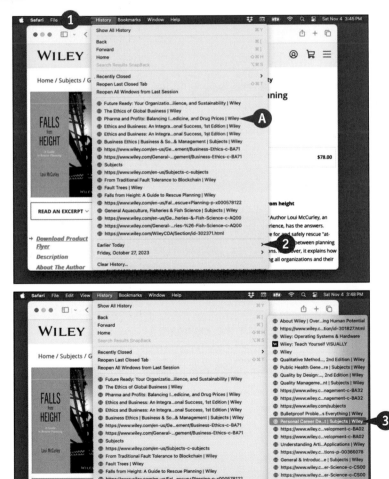

Clear Your Browsing History

1 Click **History**.

The History menu opens.

Note: You may have to scroll down the History menu to reach the Clear History command.

2 Click **Clear History**.

The Clearing History Will Remove Related Cookies and Other Website Data dialog opens.

3 Click **Clear** (⬍) and then click **Last Hour**, **Today**, **Today and Yesterday**, or **All History** to specify what you want to clear.

4 Click **Clear History**.

Safari clears the History list for the period you chose.

Note: If you want to browse without History storing the list of web pages you visit, click **File** and then click **New Private Window**. Any sites you browse in the Private Browsing window will not be stored.

TIP

What does the Show All History command do?

Click **History** and select **Show All History** to open a History window for browsing and searching the sites you have visited. Type a term in the Search box in the upper-right corner to search. Click the disclosure triangle (❯ changes to ❮) to expand the list of entries in a day. Double-click a history item to open its page in the same window, or press Control +click a history item and click **Open in New Tab** or **Open in New Window** to open the page in a new tab or a new window.

Play Music and Videos on the Web

Many websites contain music files or video files that you can play directly in the Safari browser. Safari can play many widely used types of audio files and video files, and most sites provide easy-to-use buttons — such as a Play/Pause button and a volume slider — to enable you to control playback. This section shows the SoundCloud music website and the YouTube video website as examples.

You can play music and videos either directly on your MacBook or via AirPlay to a supported device, such as an Apple TV.

Play Music and Videos on the Web

Play Music

1 In Safari, navigate to a music website and browse to find a song you want to play.

Note: This example uses the SoundCloud site, www.soundcloud.com. There are many other music sites.

2 Click **Play** (such as ▶).

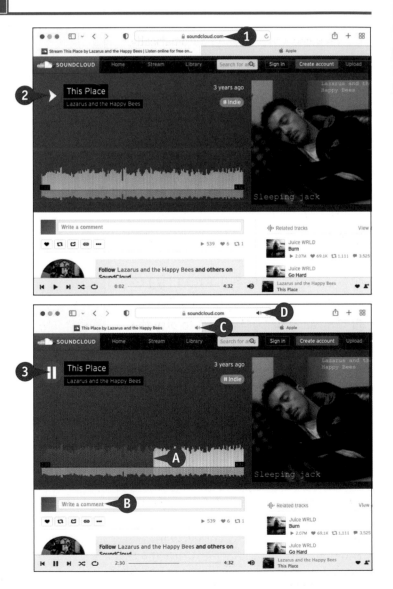

The song starts playing.

A The progress indicator shows playback progress.

3 Click **Pause** (such as ⏸) if you want to pause the music.

B Some sites enable you to comment on the music.

C The Audio icon (◀⑴) appears on a tab that is playing audio. Click **Audio** (◀⑴ changes to ◀) to mute the audio.

D If the audio is on the active tab, you can click **Audio** (◀⑴ changes to ◀) to mute the audio.

Play Videos

1 In Safari, navigate to a video website and browse to find a video you want to play.

Note: This example uses the YouTube website, www.youtube.com. There are various other video websites; see the tip for information on some.

2 Click the video to open it (not shown).

The video starts playing automatically.

E You can click **Settings** (⚙️) to change the playback speed or the quality.

3 Click **Full Screen** (▮).

The video appears full screen.

4 Move the pointer over the video.

The playback controls appear, and you can use them to control playback.

F You can click **Share** (➡️) to share the video with others.

G You can click **Exit Full Screen** (▮) to return from full screen to a window.

H You can click **AirPlay** (▮) to play the video to a TV connected to an Apple TV.

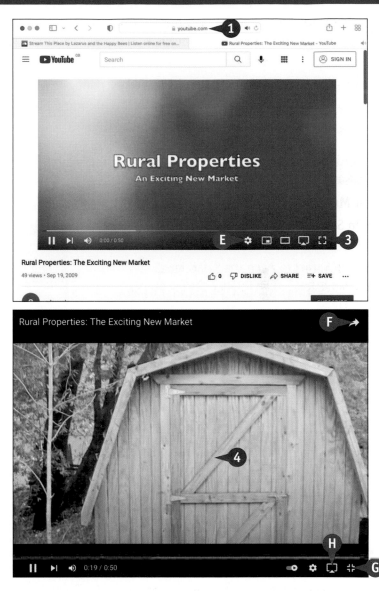

TIPS

What are the main video sites on the Web?
As of this writing, the most popular video websites include YouTube (www.youtube.com), Netflix (www.netflix.com), TikTok (www.tiktok.com), Vimeo (www.vimeo.com), and Dailymotion (www.dailymotion.com). Alternatively, search for videos using a search engine, such as Google's Videos search filter.

How do I prevent a site from playing videos automatically?
Go to the site, click **Safari** on the menu bar, and then click **Settings for** *website*, where *website* is the website's name. In the pop-up panel that opens, click **Auto-Play** (⬍) and then click **Stop Media with Sound** or **Never Auto-Play**, as appropriate.

Set Your Home Page and Search Engine

When you open a new window, Safari automatically displays your *home page*, the page it is configured to show at first. You can set your home page to any web page you want or to an empty page. You can control what Safari shows when you open a new tab or a new window: the Favorites screen, your home page, an empty page, or the page from which you opened the new tab or window. You can also specify which search engine Safari uses and configure settings for it.

Set Your Home Page and Search Engine

1 In Safari, navigate to the web page you want to make your home page.

2 Click **Safari** to open the Safari menu.

3 Click **Settings** to open the Settings window.

4 Click **General** (⚙) to display the General pane.

5 Click **Set to Current Page**.

Safari changes the Home Page text field to show the page you chose.

6 Click **New windows open with** (◉) and click **Start Page**, **Homepage**, **Empty Page**, or **Same Page**, as appropriate.

7 Click **New tabs open with** (◉) and select **Start Page**, **Homepage**, **Empty Page**, or **Same Page**, as appropriate.

8 Click **Search** (🔍).

The Search tab appears.

9 Click **Search engine** (◉) and then click the search engine you want to use.

10 Click **Private Browsing search engine** (◉) and then click the search engine for Private Browsing.

11 Select **Include search engine suggestions** (☑) to have Safari show suggestions from the search engine.

12 Select **Include Safari Suggestions** (☑) to include Safari suggestions in the Smart Search Field.

13 Select **Preload Top Hit in the background** (☑) to have Safari preload the Top Hit for the search to enable you to view it more quickly.

14 Select **Show Favorites** (☑) to include your Favorites in the Smart Search Field.

15 Select **Enable Quick Website Search** (☑) to enable the Quick Website Search feature.

16 Click **Manage Websites**.

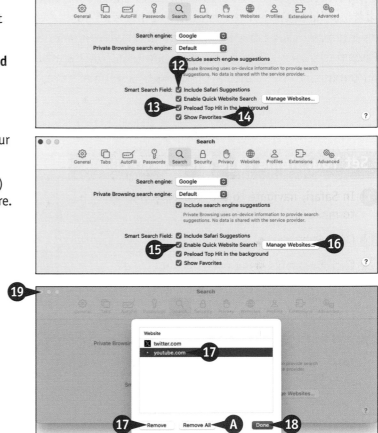

The Manage Websites dialog opens.

17 To remove a website, click it, and then click **Remove**.

A To remove all websites, click **Remove All**.

18 Click **Done**.

The Manage Websites dialog closes.

19 Click **Close** (●).

The Settings window closes.

TIPS

How can I display my home page in the current window?

Click **History** on the menu bar and then click **Home**. Alternatively, press ⌘+Shift+H.

What does Quick Website Search do?

Quick Website Search enables you to search quickly for content in some websites you have previously visited. For example, if you have previously visited the wikipedia.org web encyclopedia, you can type **wikipedia** and a search term in the address box to search for that term on Wikipedia.

Create Bookmarks for Web Pages

Safari enables you to create markers called *bookmarks* for the addresses of web pages you want to be able to revisit easily. When you find such a web page, you can create a bookmark for its address, assign the bookmark a descriptive name, and store it on the Favorites bar, on the Bookmarks menu, or in a Bookmark folder. You can then return to the web page's address by clicking its bookmark. The content of the web page may have changed by the time you return.

Create Bookmarks for Web Pages

Create a New Bookmark

1 In Safari, navigate to a web page you want to bookmark.

2 Click **Bookmarks**.

The Bookmarks menu opens.

3 Click **Add Bookmark**.

Note: You can also press ⌘+D to open the Add Bookmark dialog.

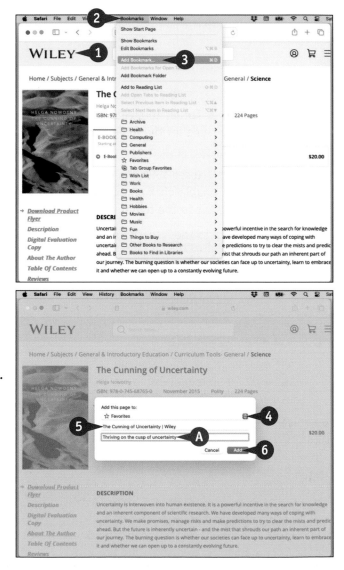

The Add Bookmark dialog opens, with the web page's title added to the text box in the middle of the dialog.

4 Click **Add this page to** (⬍) and select the location or folder in which to store the bookmark.

5 Edit the default bookmark name — the web page's title — as needed to create a descriptive name.

Ⓐ You can click **Description** and type a description to help you identify the bookmark by searching.

6 Click **Add**.

The Add Bookmark dialog closes.

Safari creates the bookmark.

Organize Your Bookmarks

1 Click **Bookmarks**.

The Bookmarks menu opens.

2 Click **Edit Bookmarks**.

Note: You can also press ⌘ + Option + B to display the Bookmarks screen.

The Bookmarks screen appears.

A You can click **New Folder** to create a new folder, type the name for the folder, and then press Return.

3 Double-click a collapsed folder to display its contents, or double-click an expanded folder to collapse it to the folder.

4 Drag a bookmark to the folder in which you want to place it.

Note: You can drag the bookmark folders into a different order. You can also place one folder inside another folder.

5 Click **Previous Page** (<).

Safari returns you to the page you were viewing before you displayed the Bookmarks screen.

TIP

How do I go to a bookmark I have created?

If you placed the bookmark on the Favorites bar, click the bookmark on the Favorites bar; if the Favorites bar is not displayed, click **View** and **Show Favorites Bar** to display it. If you put the bookmark on the Bookmarks menu, click **Bookmarks**, and then click the bookmark on the Bookmarks menu or one of its continuation menus. If you cannot easily locate the bookmark, click **Show sidebar** (□) to display the sidebar, and then click **Bookmarks** (🔖) to display the bookmarks. Locate the bookmark, and then double-click it.

Using Reader View and Reading List

Safari's Reader View enables you to minimize distractions by displaying only the text of a web page in an easily readable format. Reader View works on many, but not all, web pages.

Safari also enables you to save a web page in its current state so that you can read it later. You can quickly add the current web page to Reading List, which you can then access via the Reading List pane in the sidebar.

Using Reader View and Reading List

Switch a Web Page to Reader View

1 In Safari, navigate to a web page you want to read.

2 Click **Show Reader View** (⊟).

Note: If the Show Reader View button does not appear, it is because Safari has determined that the page does not have content suitably formatted for Reader View.

Note: You can also switch to Reader View by clicking **View** and **Show Reader** and switch back by clicking **View** and **Hide Reader**. Alternatively, press ⌘+Shift+R to toggle Reader View on or off.

Safari switches the web page to Reader View.

A The web page's contents appear in easy-to-read fonts.

B Links are easier to see but work as usual.

3 When you finish using Reader View, click **Hide Reader View** (⊟).

Safari switches the web page back from Reader View.

Add a Web Page to Reading List

1 Navigate to a web page.

2 Move the pointer over the address box.

3 Click **Add page to Reading List** (⊕).

Safari displays a brief animation showing the page moving from the address box to the Show Sidebar button (▢).

Note: You can also click **Bookmarks** and select **Add to Reading List** or press ⌘+Shift+D.

Note: You can click **Bookmarks** and select **Add These Tabs to Reading List** to add all the tabs in the current window to Reading List.

Open Reading List and Display a Page

1 Click **Show sidebar** (▢).

Safari displays the sidebar.

2 Click **Reading List** (∞) (not shown).

The Reading List pane opens.

3 Click the item you want to read.

The item appears.

ⓒ When you are ready to remove an item from Reading List, press Control+click its entry and click **Remove Item**.

TIP

Can I make a site always open in Reader View?
Yes — when Reader View is available for the site's pages. Go to the site, click **Safari** on the menu bar, and then click **Settings for This Website**. In the pop-up panel that opens, select **Use Reader when available** (☑).

Organize Safari Tabs into Tab Groups

I f you browse extensively in Safari, you will likely open many tabs. You can organize your tabs to some extent by devoting a window to each major topic you are exploring and dragging tabs from one window to another as needed. But Safari also provides a better way to organize your tabs: tab groups.

As you open each new tab, Safari adds it to the ungrouped tabs. When you want to start organizing, you can create one or more tab groups, shuffle your open tabs into the groups, and display the group you want to work with.

Organize Safari Tabs into Tab Groups

1 Open various tabs in a single Safari window.

2 Click **Tab Group Picker** (⌄).

The Tab Group Picker pop-up menu opens.

Ⓐ If you want to put all your tabs in a single group, click **New Tab Group with *N* Tabs**.

3 Click **New Empty Tab Group**.

Safari opens a new tab.

The sidebar appears.

Safari creates a new tab group with the provisional name *Untitled* and selects this name for you to type over.

4 Type the name for the tab group and press Return.

Safari applies the name to the tab group.

5 Create further tab groups by clicking **New tab group**, clicking **New Empty Tab Group**, typing a name, and pressing Return.

6 Click your list of ungrouped tabs.

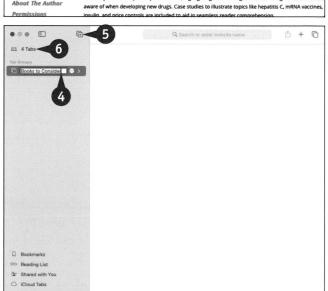

The ungrouped tabs appear.

⑦ Press `Control`+click the tab you want to assign to a group.

The contextual menu opens.

⑧ Click or highlight **Move to Tab Group**.

The continuation menu opens.

⑨ Click the appropriate tab group.

Safari adds the tab to the group.

⑩ Repeat steps **7** to **9** to assign your tabs to groups (not shown).

⑪ Move the pointer over the tab group you want to view.

The More icon (⬤) appears.

⑫ Click **More** (⬤).

The More menu appears.

⑬ Click **Show Tab Overview**.

The tab overview appears, showing a thumbnail of each tab.

⑭ Click the tab you want to view.

Ⓑ You can click Add (➕) to add a new tab to the group.

The tab appears.

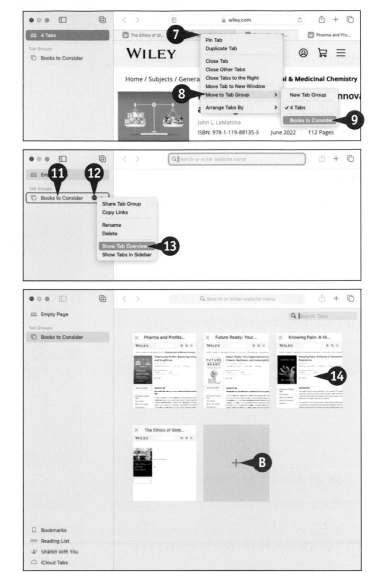

TIPS

What other way can I assign a tab to a tab group?
You can drag the tab from the tab bar at the top of the Safari window to the appropriate tab group in the Tab Groups list in the sidebar.

How do I delete a tab group?
Press `Control`+click the tab group in the Tab Groups list in the sidebar, and then click **Delete** on the contextual menu.

Download a File

M any websites provide files to download, and Safari makes it easy to download files from websites to your MacBook's file system. For example, you can download apps to install on your MacBook, pictures to view on it, or songs to play.

macOS includes apps that can open many file types, including music, graphic, movie, document, and PDF files. To open other file types, you may need to install extra apps or add plug-in software components to extend the features of the apps you already have.

Download a File

1 In Safari, go to the web page that contains the link for the file you want to download.

2 Click the link.

Note: For safety, download files only from sites that you trust, and use an anti-malware app or antivirus app to scan downloaded files before you open them.

3 If the Do You Want to Allow Downloads? dialog opens, click **Allow**.

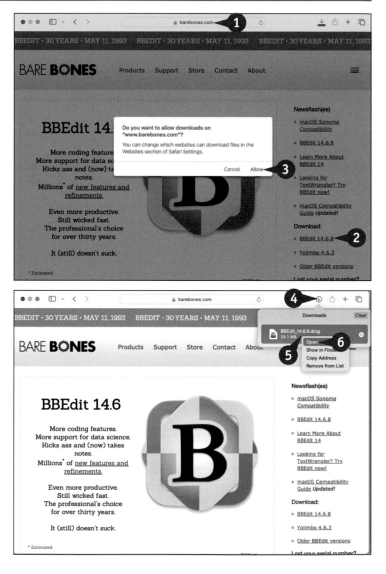

Safari starts the download.

Note: The indicator on the Downloads button ( ) shows the progress of the download.

4 Click **Downloads** ( ) to open the Downloads panel.

5 When the download is complete, press +click the file in the Downloads panel.

The contextual menu opens.

6 Click **Open**.

Note: Depending on the file type and the settings you have chosen, Safari may open the file automatically for you.

The file opens.

Depending on the file type, you can then work with the file, enjoy its contents, or install it.

Note: If the file is an app, you can install it, as discussed in Chapter 4. If the file is a text file, such as a document or a picture, macOS opens the file in the app for that file type.

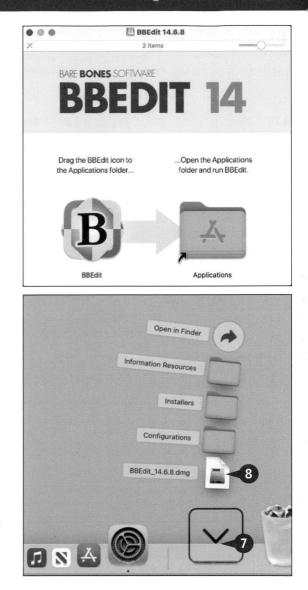

7 If the file disappears from the Downloads window and Safari does not open the file for you, click **Downloads** on the Dock.

The Downloads stack opens.

8 Click the file you downloaded.

The file opens.

TIP

What should I do when clicking a download link opens the file instead of downloading it?
If clicking a download link on a web page opens the file instead of downloading it, press Control + click the link to display the contextual menu. Click **Download Linked File** if you want to save the file in your Downloads folder under the file's current name. If you prefer to save the file in a different folder or under a different name of your choice, click **Download Linked File As**, and then specify the folder and filename in the dialog that opens.

Create and Use Safari Profiles

If you need to keep different areas of your Safari browsing separate, you can use Safari's profiles. For example, you might create a Work profile that you could use for your work-related browsing, keeping it separate from your home or leisure browsing. The profile contains Safari information including your browsing history, bookmarks, and tab groups.

When you create a new profile, Safari automatically creates a Personal profile for you — so after creating one profile, you have two profiles. Given this, you would normally create a non-personal profile, such as a Work profile or a School profile, rather than a Home profile.

Create and Configure a New Profile

To create a new profile, click **Safari** on the menu bar, and then click **Create Profile**. Safari displays the Profiles tab of the Settings window. At first, this tab contains only information about profiles and the Start Using Profiles button.

Click **Start Using Profiles** (A) to open a dialog for creating a new profile.

In the Name box (B), type the name for the profile, such as **Work**.

On the Symbol line (C), click the symbol to use; or click **More** (D, ⋯) to display a pop-up panel with many symbols, and then click a symbol.

On the Color line (E), click the swatch (such as ■), and then click the color for the profile.

In the Favorites area (F), select **Create new bookmarks folder** (G, ●) to have Safari create a new bookmarks folder for the profile; this is often the best choice. However, if an existing bookmarks folder already contains bookmarks you want to use in the profile, select **Use existing folder** (H, ●), click the pop-up menu (I, ⬍), and then select the folder.

When you have made your choices, click **Create Profile** (J). The dialog closes, and Safari creates the profile.

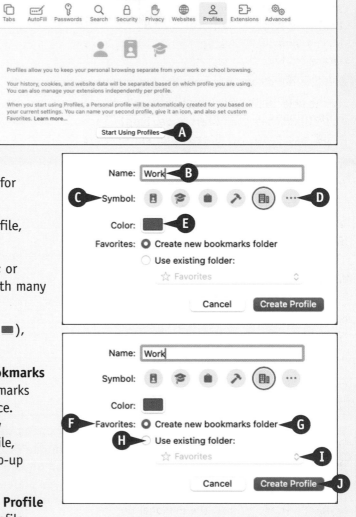

194

The Profiles tab of the Settings window displays the profile you have created (K), the Personal profile that Safari has now automatically created and made the default (L), and controls for managing profiles.

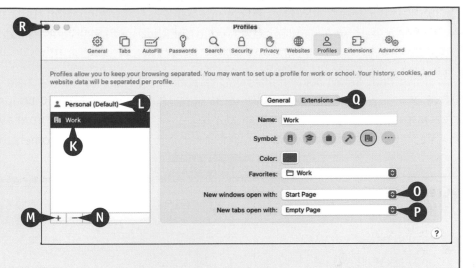

You can add a new profile by clicking **Add** (M, **+**). You can delete a profile by clicking it, clicking **Remove** (N, **—**), and then clicking **Delete** in the confirmation dialog that opens.

For the selected profile, you can choose how new windows and tabs open. Click **New windows open with** (O, ⬍) and then click **Start Page**, **Empty Page**, **Same Page**, or **Tabs for** *profile*. Click **New tabs open with** (P, ⬍) and then click **Start Page**, **Empty Page**, or **Same Page**.

You can also click **Extensions** (Q) to display the Extensions subtab, on which you can configure Safari extensions for use in the profile. An *extension* is add-on software that provides extra functionality, such as an ad blocker.

When you finish configuring new profiles, click the profile you want to make active, and then click **Close** (R, ●) to close the Settings window.

Work in a Profile and Switch Profiles

Once you have set up profiles, the Profiles button appears on the toolbar. In an active window, the Profiles button (S) shows the color you assigned to the profile. In an inactive window, the Profiles button (T) appears in gray.

Once you have opened a window for the profile you want to use, you can work in the window as normal. When you open a new tab in the window, the tab belongs to that profile.

To switch to another profile, click **Profiles** (such as **Work**, S) in the active window, and then click the appropriate New Window command on the pop-up menu. For example, you might click **New Personal Window** (U) to open a window using the Personal profile.

Configure Safari for Security and Privacy

The Web is packed with fascinating sites and useful information, but it is also full of malefactors and criminals who want to attack your MacBook and steal your valuable data.

To keep your MacBook safe, you can prevent Safari from automatically opening supposedly safe files you download. Safari also enables you to choose whether to enable JavaScript, whether to allow cross-site tracking, and whether to hide your MacBook's IP address. You can also choose which Internet plug-ins to use and which websites can use them.

Configure Safari for Security and Privacy

1 In Safari, click **Safari** to open the Safari menu.

2 Click **Settings** to open the Settings window.

3 Click **General** (⚙️) to display the General pane.

4 Deselect **Open "safe" files after downloading** (⬜).

5 Click **Security** (🔒).

The Security pane appears.

6 Select **Warn when visiting a fraudulent website** (✅).

7 Deselect **Enable JavaScript** (⬜) if you want to disable JavaScript. See the tip for advice.

8 Click **Privacy** (🖐️).

The Privacy pane appears.

9 Select **Prevent cross-site tracking** (✅) to have Safari try to prevent sites tracking your movements from one site to another.

10 Select **Hide IP address from trackers** (✅) to have Safari hide your computer's Internet address from known trackers.

11 Select **Require Touch ID to view locked tabs** (✅) to have macOS demand your fingerprint to unlock locked Private Browsing tabs.

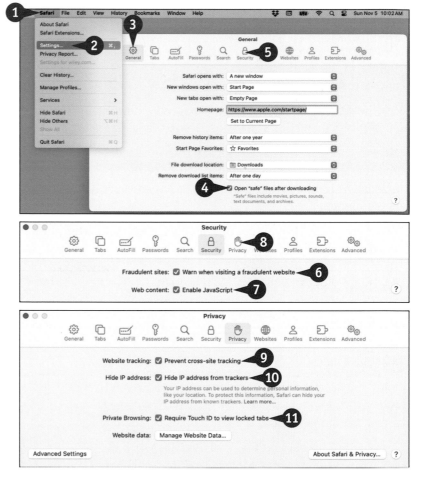

A If your MacBook does not have Touch ID, the Require Password to View Locked Tabs check box appears instead of the Require Touch ID to View Locked Tabs check box.

12 Click **Manage Website Data**.

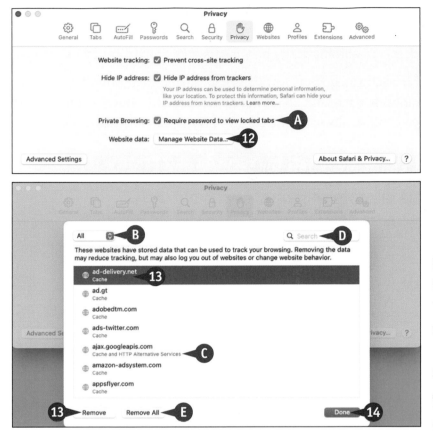

The Manage Website Data dialog opens.

B If you have configured Safari profiles, this pop-up menu appears in the upper-left corner. Click 🔃, and then either click **All** to affect all profiles or click the profile you want to affect.

C Each entry lists the types of data the site has stored about you.

D You can search to locate a site.

13 To remove the data a website has saved about you, click the site, and then click **Remove**.

E You can click **Remove All** to remove all the sites' saved data.

14 Click **Done**.

The Manage Website Data dialog closes.

The Privacy tab of the Settings dialog box appears again.

TIP

What is JavaScript, and should I disable it?

JavaScript is a scripting language used by many websites to provide interactive features. While it is possible for malefactors to perform some unwelcome actions with JavaScript, such as borrowing your computer's processing power to "mine" cryptocurrency, the language is widely used for positive purposes and is required for many websites to be usable, so you should not disable it. What is more of a threat to your MacBook's security is Java, a full-featured programming language that can run on many computer platforms and is sometimes used by web-delivered services. JavaScript is completely different from Java but is sometimes tainted by association with Java. Safari has no setting for enabling or disabling Java.

continued ▶

In the Websites pane of Safari settings, you can configure various aspects of how Safari interacts with websites you visit for security, privacy, and usability. For example, you can control which sites can automatically play media and which sites can send you notifications.

For each category of settings, you can choose custom settings for individual websites you visit and a global setting that applies to all websites for which you have not chosen custom settings. For example, you might allow specific sites to play videos automatically but block all other websites from doing so.

Configure Safari for Security and Privacy (continued)

15 Click **Websites** (⊕).

16 In the General pane, click the category you want to customize first. To follow this example, click **Auto-Play** (▶).

17 In the Currently Open Websites list, click ◆ for a site you want to configure, and then click **Allow All Auto-Play**, **Stop Media with Sound**, or **Never Auto-Play**, as needed.

18 Adjust the settings for sites in the Configured Websites list, as needed.

19 Click **When visiting other websites** (◆) and click the setting to use for all other websites.

Ⓐ You can click a website in the Configured Websites list and then click **Remove** to remove it.

20 Click **Downloads** (⊙).

21 In the Currently Open Websites list, click ◆ for a site you want to configure, and then click **Ask**, **Deny**, or **Allow**, as needed.

22 Adjust the settings for sites in the Configured Websites list, as needed.

23 Click **When visiting other websites** (◆) and click the setting to use for all other websites.

24 Click **Notifications** (◻).

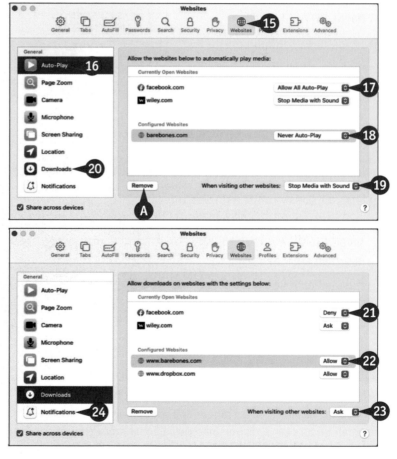

25 In the These Websites Have Asked for Permission to Show Alerts in Notification Center list, click 🔁 for a site you want to configure, and then click **Allow** or **Deny**, as needed.

26 Select (☑) or deselect (☐) **Allow websites to ask for permission to send notifications**, as needed.

27 Select (☑) or deselect (☐) **Share across devices** to control whether Safari shares your MacBook's settings with the other devices that log into your iCloud account.

28 Click other categories and choose settings, as needed. For example, click **Camera** (📷) and specify which websites can use your MacBook's camera.

29 Click **AutoFill** (🖊).

30 In the AutoFill Web Forms list, click to select (☑) or deselect (☐) **Using information from my contacts**, **User names and passwords**, **Credit cards**, and **Other forms**, to specify the items for which you want to use AutoFill.

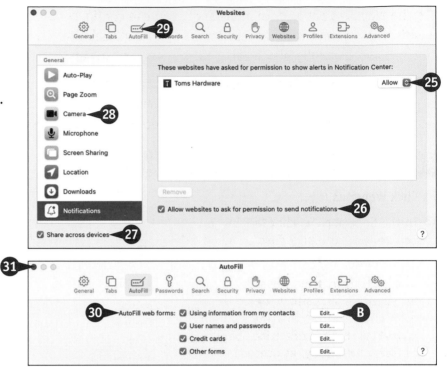

B You can click **Edit** to view the stored information.

Note: Clicking **Edit** for Using Information from My Contacts opens the Contacts app.

31 Click **Close** (⬤).

The Safari Settings window closes.

TIP

What are cookies, and should I block them?

A *cookie* is a small text file that a website uses to store information about what you do on the site — for example, what products you have browsed or added to your shopping cart. Cookies from sites you visit are usually helpful to you. Be aware that blocking all cookies may prevent many websites from working as they are supposed to.

CHAPTER 7

Sending and Receiving E-Mail

macOS includes Mail, a powerful but easy-to-use e-mail app. After setting up your e-mail accounts, you can send and receive e-mail messages and files.

Set Up Your E-Mail Accounts

The Mail app enables you to send and receive e-mail messages easily using your existing e-mail accounts. If you add an iCloud account to your user account during initial setup or in System Settings, macOS can automatically set up e-mail in Mail. After setup, you can add other e-mail accounts manually. For some accounts, you need only your e-mail address and your password. For other accounts, you also need to enter the addresses and types of your provider's mail servers. If the provider offers two-factor authentication, which gives you an extra layer of security, you should use it.

Set Up Your E-Mail Accounts

1 Click **Mail** (📧) on the Dock to open Mail.

2 Click **Mail** to open the Mail menu.

3 Click **Add Account** to open the Choose a Mail Account Provider dialog.

4 Click the account type (○ changes to ◉). Your choices are **iCloud**, **Microsoft Exchange**, **Google**, **Yahoo!**, **AOL**, or **Other Mail Account**.

5 Click **Continue**.

A dialog opens that allows you to sign in to the account. This example shows the Sign In dialog that appears for a Google account.

6 Type your e-mail address.

7 Click **Next**, type your password when prompted, and then click **Next** again.

If you use two-factor authentication for the account, a dialog opens prompting you to provide the second form of authentication — for example, a code texted to your phone.

8 Enter the code.

9 Select **Don't ask again on this device** (☑) if appropriate.

10 Click **Next**.

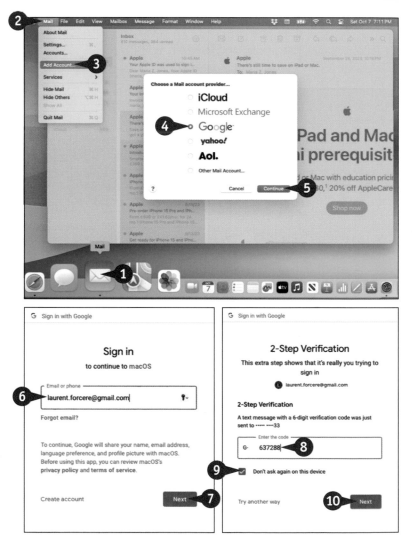

11 If a dialog such as the macOS Wants to Access Your Google Account dialog opens, read its details, and then click **Allow** if you want to proceed.

The Select the Apps You Want to Use with This Account dialog opens.

12 Select the check box (☑) for each app you want to use with this account.

13 Click **Done**.

Mail displays your inbox, and you can begin reading and sending e-mail.

A You can click a message in the message list to display its contents.

B Mail automatically identifies possible events and contacts in e-mail messages you receive.

C You can click **Add** to add the event to your calendar.

D You can click **Close** (✕) to close the panel without adding the event.

Note: When you receive a calendar invitation, the message includes the name of the event and the Calendar icon. You can click **Accept**, **Maybe**, or **Decline** to give your response.

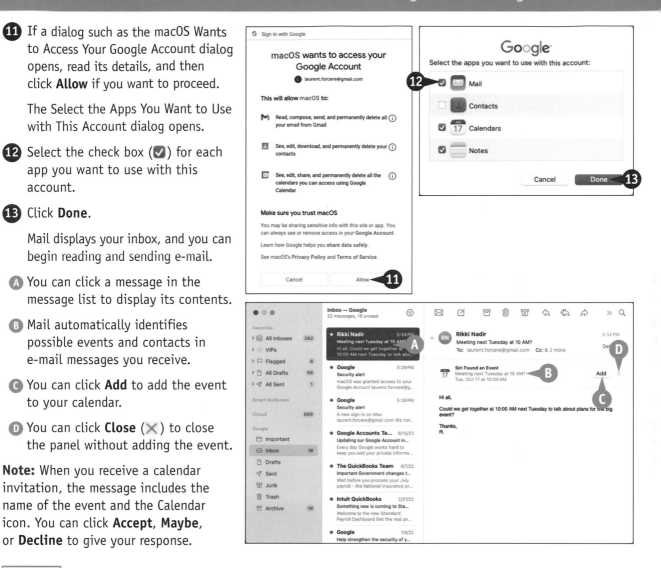

TIPS

How do I add a Microsoft 365 account?

Click **Microsoft Exchange** (◯ changes to ◉) in the Choose a Mail Account Provider dialog. Use the server name outlook.office365.com.

What extra information must I provide when I select Other Mail Account?

You must choose between IMAP and POP for your incoming mail server type and enter the mail server's address. Check your ISP's website or call customer service to find out what type of mail server to use. Most ISPs use either POP (Post Office Protocol) or IMAP (Internet Mail Access Protocol), but others use Exchange or Exchange IMAP. You may also need to provide the outgoing mail server's address.

Send an E-Mail Message

The Mail app enables you to send an e-mail message to anybody whose e-mail address you know. You can specify the recipient's address either by typing it directly into the To field or by selecting it from your contacts list.

You can send an e-mail message to a single person or to multiple people. You can send copies to Cc, or carbon-copy, recipients or send hidden copies to Bcc, or blind carbon-copy, recipients. After you click the Send button, Mail briefly delays sending the e-mail to allow you to undo the action if you realize something is wrong.

Send an E-Mail Message

① In Mail, click **New Message** (✏️) or press ⌘+N.

Note: To display text labels for the buttons on the toolbar in either the main Mail window or a message window, press Control+click or right-click the toolbar, and then click **Icon and Text** on the contextual menu.

Ⓐ A new message window opens.

② Click **Add Contact** (⊕).

③ Click the contact.

④ Click the e-mail address.

Note: In the Contacts panel, a name in lighter gray has no e-mail address in the contact record.

The contact name appears as a button in the To field.

Note: You can add other addresses to the To field as needed.

⑤ To add a Cc recipient, click the Cc field.

⑥ Start typing a name or e-mail address.

Mail displays matches from Contacts.

⑦ Click the appropriate match or finish typing.

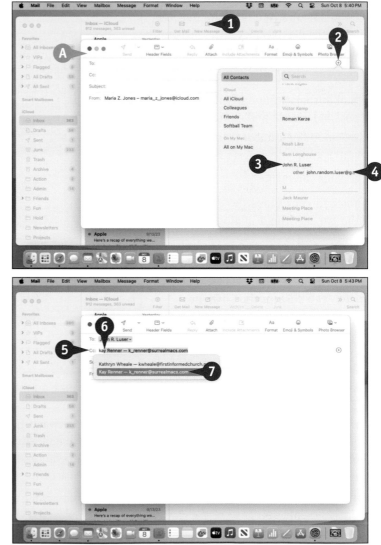

The name or address appears in the Cc box.

8 Type the subject for the message.

9 Type the body text of the message.

B You can click **Show Format Bar** (Aa) to display the Format bar, which contains controls for setting the font family, font size, and font color, and applying boldface, italics, or underline. For more formatting choices, press ⌘+T to open the Fonts window.

Note: You can switch a message between plain text and rich text by clicking **Format** on the menu bar and then clicking **Make Rich Text** or **Make Plain Text**.

10 Click **Send** (◁).

The message window closes.

C The Undo Send button appears in the main Mail window. You can click this button to stop sending the message and reopen it for editing.

Note: The Undo Send button appears for 10 seconds by default. To change the delay, click **Mail**, click **Settings**, click **Composing** (◻), click the **Undo Send Delay** pop-up menu (◆), and then select **20 Seconds**, **30 Seconds**, or **Off**.

Mail sends the message and stores a copy in your Sent folder.

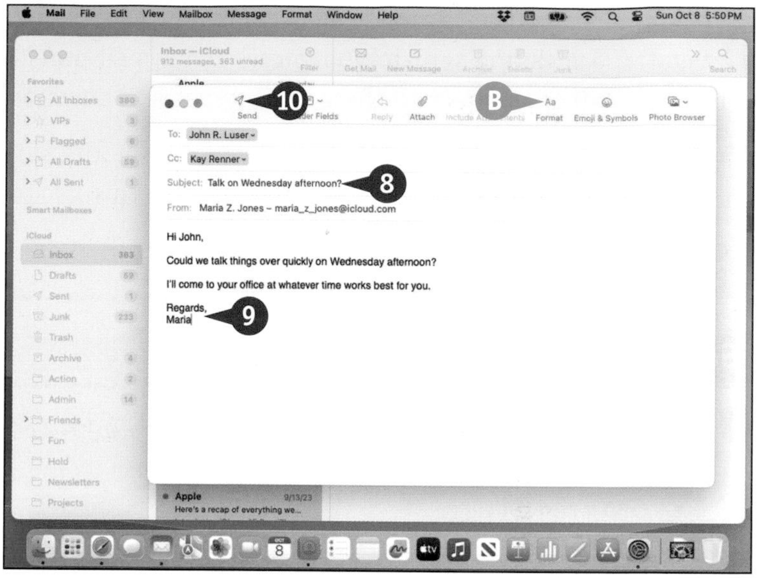

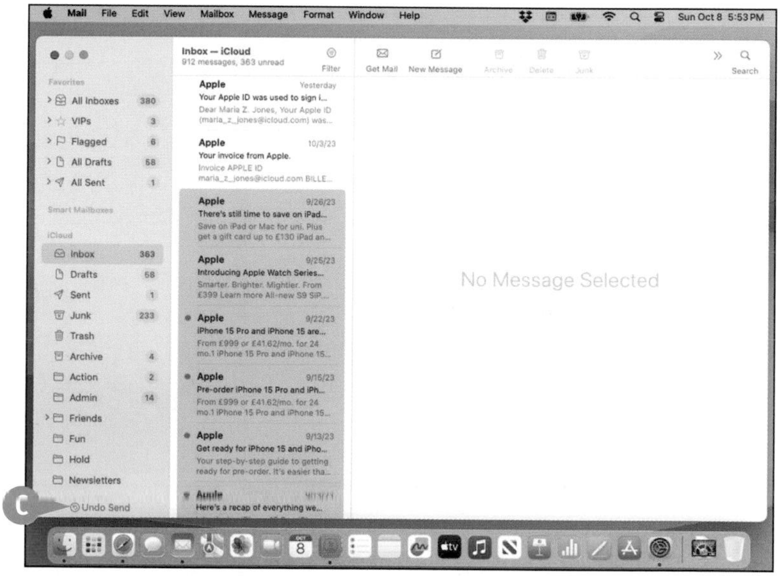

TIP

How can I send Bcc (blind carbon-copy) messages?

Mail hides the Bcc field in the New Message window by default. To display the Bcc field, click **Header Fields** (◻⌄) on the toolbar of a message window and then click **Bcc Address Field** on the pop-up menu. The Bcc field appears below the Cc field, and you can add recipients either by clicking **Add Contact** (⊕) or by typing their names or e-mail addresses.

Each Bcc recipient sees only their own address, not the addresses of other Bcc recipients. The To recipients and Cc recipients see none of the Bcc recipients' names and addresses.

Receive and Read Your Messages

Mail enables you to receive your incoming messages easily and read them in whatever order you prefer. A message sent to you goes to your mail provider's e-mail server. To receive the message, you cause Mail to connect to the e-mail server and download the message to your MacBook.

When working with e-mail, it often helps to display the sidebar, which contains the list of mailboxes, on the left side of the Mail window. You can use this list to navigate among mailboxes.

Receive and Read Your Messages

Display the Mailboxes List and Receive Messages

1 In Mail, if the sidebar is hidden, click **View**.

2 Click **Show Sidebar**.

Note: You can also press ⌘ + Control + S to toggle the display of the sidebar.

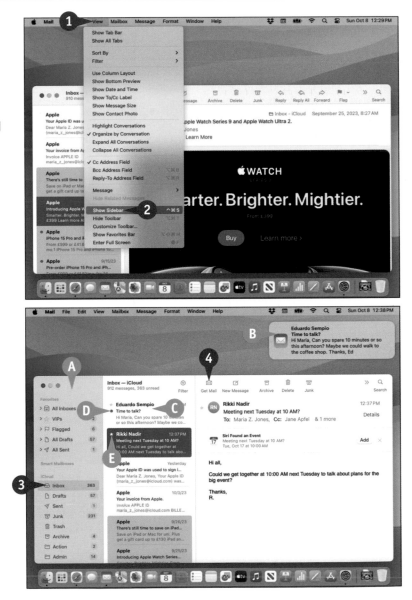

A The Sidebar appears.

3 Click the mailbox you want to view.

4 Click **Get Mail** (✉).

Mail connects to the e-mail server and downloads any messages.

B Depending on how you have configured notifications, a notification banner or alert may appear as a message arrives.

C The new messages appear in your inbox.

D A blue dot (●) indicates an unread message.

E A star (★) indicates a message from someone you have designated a VIP.

Read Your Messages

1 Click a message in the message list.

F The message appears in the reading pane.

2 Double-click a message in the message list.

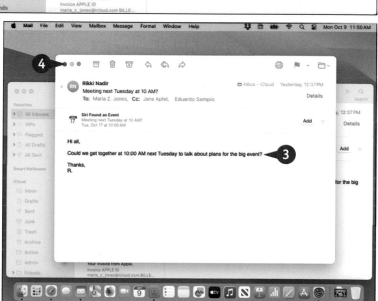

The message's text and contents appear in a separate window.

3 Read the message.

4 Click **Close** (●).

The message window closes.

Note: Opening a message in a separate window can be helpful for focusing on the message. Opening multiple messages in separate windows enables you to compare the messages easily.

Is there an easy way to tell whether I have new messages?

If you have unread messages, the Mail icon on the Dock shows a red circle (such as ①) or a rounded rectangle (such as ③⑦④) containing the number of unread messages.

How can I change Mail's frequency of checking for new messages?

Click **Mail** and **Settings** to open the Settings window, and then click **General** (⚙). Click the **Check for new messages** pop-up menu (◉), and then select the method or interval: **Automatically**, **Every minute**, **Every 5 minutes**, **Every 15 minutes**, **Every 30 minutes**, **Every hour**, or **Manually**. Click **Close** (●) to close the Settings window.

Reply to a Message

Mail enables you to reply to any e-mail message you receive. When you reply, you can include either the entire original message or just the part of it that you select.

If you are one of multiple recipients of the message, you can choose between replying only to the sender and replying to both the sender and all the other recipients other than Bcc recipients. You can also adjust the list of recipients manually if necessary, removing existing recipients and adding other recipients.

Reply to a Message

1 In the inbox, click the message to which you want to reply.

Note: You can also double-click the message to open it in a message window, and then start the reply from there.

2 Click **Reply** (↩).

Note: If the message has multiple recipients, you can click **Reply All** (↩↩) to reply to the sender and to all the other recipients except Bcc recipients.

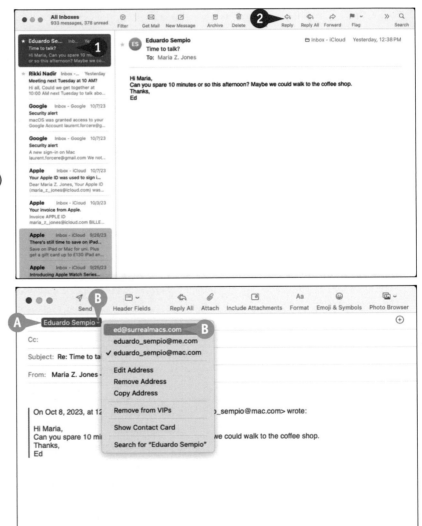

Mail creates the reply and opens it in a window.

A The recipient's name appears as a button.

B You can position the pointer over the contact's name, click the pop-up button (⏷) that appears, and then click a different address if necessary.

3 Type the text of your reply.

It is usually best to type your text at the beginning of the reply rather than after the message you are replying to.

Note: You can also add other recipients to the message as needed. If you have chosen to reply to all recipients, you can remove any recipients as necessary.

4 Click **Send** ().

Mail sends the reply and saves a copy in your Sent folder.

Note: Click **Message** and then click **Send Again** to send the same message again — for example, because the recipient has deleted it by accident. You can change recipients or the message contents as needed.

TIP

How do I reply to only part of a message rather than send the whole of it?
Select the part you want to include, and then click **Reply** (↩) or **Reply All** (↩), as appropriate. Mail creates a reply containing only the part you selected. If Mail still includes all of the message, click **Mail**, click **Settings**, click **Composing** (✐), and then click **Include selected text, if any; otherwise include all text** (◯ changes to ◉) instead of **Include all of the original message text**.

Forward a Message

Mail enables you to forward to other people a message that you receive. You can either forward the entire message or forward only a selected part of it.

When you forward a message, you can add your own comments to the message. For example, you might want to explain to the recipient which person or organization sent you the original message, why you are forwarding it, and what action — if any — you expect them to take.

Forward a Message

Ⓐ The Replied arrow (↩) indicates you have replied to a message.

Ⓑ The Forwarded arrow (↪) indicates you have forwarded a message.

① Click the message.

The message's content appears in the reading pane.

Note: You can also forward a message that you have opened in a message window.

② Click **Forward** (↪).

Note: You can click **Message** and then click **Redirect** to redirect a message to someone else without the Fwd: indicator appearing.

A window opens showing the forwarded message.

The subject line shows Fwd: and the message's original subject, so the recipient can see it was forwarded.

③ Enter the recipient's name or address. You can either type the address or click **Add Contact** (⊕) and then select the address from the Contacts panel.

④ Edit the subject line of the message if necessary.

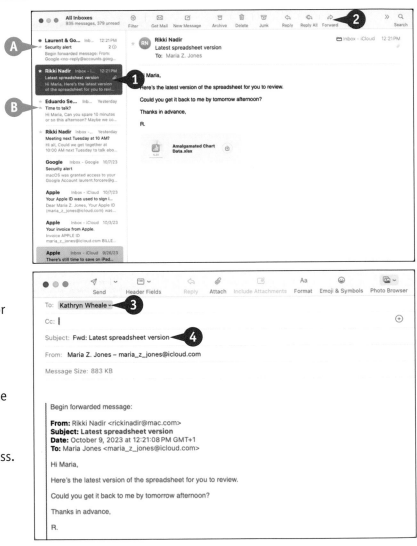

5 Optionally, edit the forwarded message to shorten it or make it clearer to the recipient.

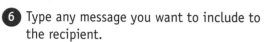

6 Type any message you want to include to the recipient.

7 Click **Send** (✈).

Mail sends the forwarded message to the recipient.

TIPS

What does the Forward as Attachment command on the Message menu do?

The Forward as Attachment command enables you to send a copy of a message as an attachment to a message instead of in the message itself. This command is useful when you want to send a forwarded message that includes formatting in a text-only message.

How do I forward only part of a message rather than all of it?

Select the part you want to forward, and then click **Forward** (↪). Mail includes only the part you selected.

Send a File via E-Mail

As well as enabling you to communicate via e-mail messages, Mail gives you an easy way to transfer files to other people. You can attach one or more files to an e-mail message so that the files travel as part of the message. The recipient can then save the file on their computer and open it.

Mail's Send Large Attachments with Mail Drop feature enables you to transfer large files without running up against the size constraints that some mail servers impose.

Send a File via E-Mail

1 In Mail, click **New Message** (⬚).

Note: If you have multiple accounts, tell Mail which account to use for outgoing messages. Click **Mail** and **Settings** to open the Settings window, and then click **Composing** (⬚). In the Addressing section of the Composing pane, click the **Send new messages from** pop-up menu (⬚), and then either click the account to use or click **Automatically select best account** to let Mail choose the account.

A new message window opens.

2 Add the recipient's name or address. You can either type the address or click **Add Contact** (⊕) and then select the address from the Contacts panel.

3 Type the subject for the message.

4 Type any message body text that is needed.

5 Click **Attach** (⬚).

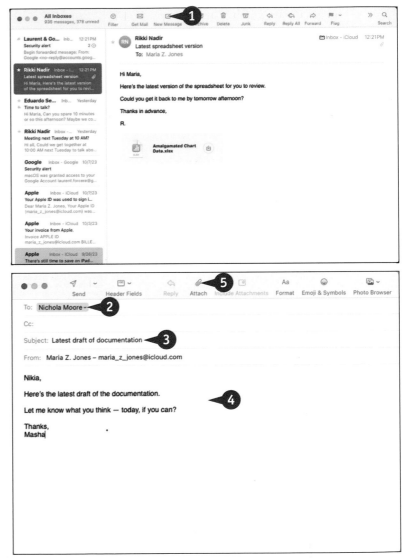

A dialog opens.

6 Click the file you want to attach to the message.

A If you know the recipient's computer runs Windows, click **Show Options** to reveal the Send Windows-Friendly Attachments check box, and then select **Send Windows-Friendly Attachments** (☑).

7 Click **Choose File**.

Note: Ask yourself whether the recipient will have an app that can open the file you are sending. The Portable Document Format, PDF, is usually a safe choice, because many people have a PDF reader installed; if not, they can download a free one.

The dialog closes.

B Mail attaches the file to the message.

Note: Depending on the file type, the attachment may appear as an icon in the message or as a picture.

C The Message Size readout appears below the Subject line.

8 Click **Send** (✐).

Mail sends the message with the file attached.

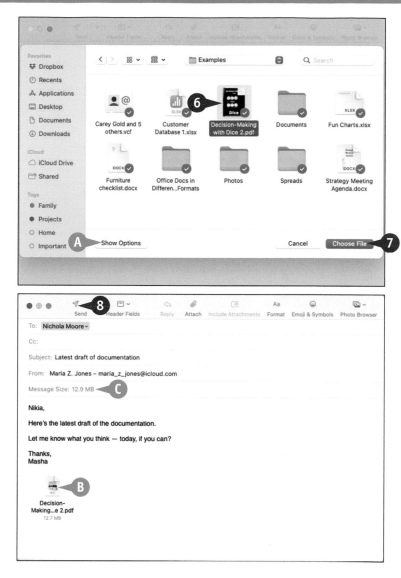

TIP

What does Send Large Attachments with Mail Drop do?

Send Large Attachments with Mail Drop uploads large attachments to Apple's Internet storage instead of including them in the message. If the message's recipient is using an iCloud account, they receive the attachments in the e-mail message via automatic download; if the recipient is using another type of account, they receive a link to download the file manually.

To enable Send Large Attachments with Mail Drop, click **Mail** on the menu bar and then click **Settings**. Click **Accounts** (@) to display the Accounts pane, click the account in the left pane, and then click **Account Information**. Select **Send large attachments with Mail Drop** (☑), and then click **Close** (●).

Receive a File via E-Mail

A file you receive via e-mail appears as an attachment to a message in your inbox. You can use the Quick Look feature to examine the file and decide whether to keep it or delete it. Quick Look can display the contents of many types of files well enough for you to determine what they contain.

To keep an attached file, you can save it to your MacBook's drive. You can then remove the attached file from the e-mail message to help keep down the size of your mail folder.

Receive a File via E-Mail

1 In your inbox, click the message.

The message appears in the reading pane.

2 Press **Control**+click the attachment.

The contextual menu opens.

3 Click **Quick Look Attachment**.

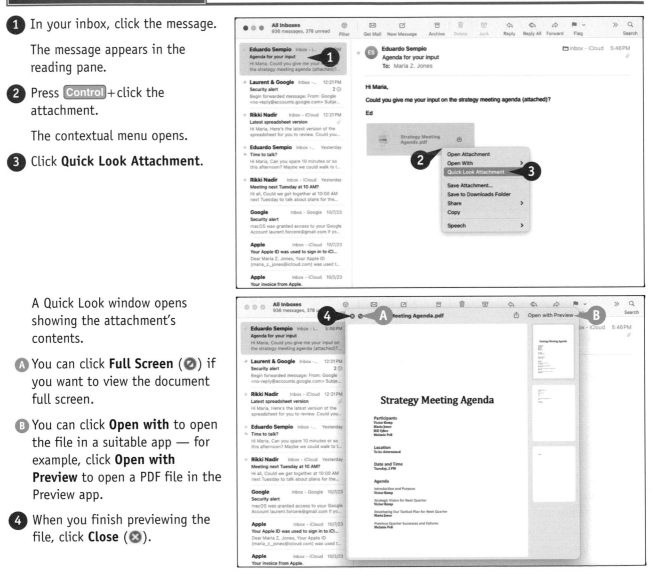

A Quick Look window opens showing the attachment's contents.

A You can click **Full Screen** (⊘) if you want to view the document full screen.

B You can click **Open with** to open the file in a suitable app — for example, click **Open with Preview** to open a PDF file in the Preview app.

4 When you finish previewing the file, click **Close** (⊗).

The Quick Look window closes.

5 Press Control+click the attachment.

The contextual menu opens.

6 Click **Save Attachment**.

The Save As dialog opens.

Note: If the Save As dialog opens at its small size, click **Expand Dialog** (⌄) to expand it.

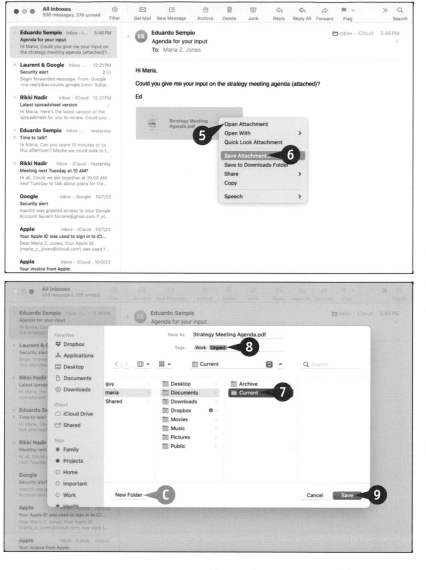

7 Navigate to the folder in which you want to save the file.

C You can click **New Folder** to create a new folder in the current folder.

8 Optionally, click the **Tags** box and click each tag you want to assign to the file.

9 Click **Save**.

Mail saves the file.

Note: If you want to remove the attachment from the message, click **Message** and then click **Remove Attachments**.

TIPS

Should I check incoming files for viruses and malevolent software?

Yes, you should always check incoming files with antivirus software. Most antivirus and security apps scan incoming files automatically, but they also enable you to scan individual files manually. Even though Macs generally have fewer problems with viruses and malevolent software than Windows PCs, it is possible for a file to cause damage, steal data, or threaten your privacy.

Is there a quick way to see what messages have attachments?

Click the appropriate mailbox, then click **View** on the menu bar, highlight **Sort By** on the View menu, and then click **Attachments**. Mail sorts the mailbox so that the messages with attachments appear first.

View E-Mail Messages by Conversations

Mail enables you to view an exchange of e-mail messages as a conversation instead of viewing each message as a separate item. Conversations, also called *threads*, let you browse and sort messages on the same subject more easily by separating them from other messages in your mailboxes. A conversation can also pull together messages contained in different mailboxes — for example, incoming messages might be in your inbox, while your replies are in the Sent mailbox.

If you decide to organize your messages by conversations, you can expand or collapse all conversations to see the messages you want.

View E-Mail Messages by Conversations

1 In Mail, click the mailbox that contains the messages you want to view.

A The messages in the mailbox appear.

B When messages are not organized by conversations, the messages in a conversation appear individually in the mailbox.

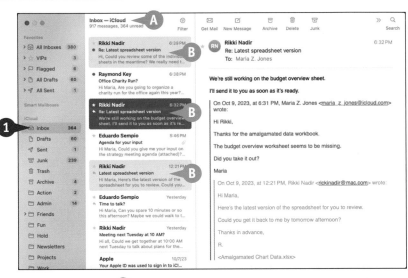

2 Click **View**.

The View menu opens.

3 Click **Organize by Conversation**.

Note: To change the order in which the conversation's messages appear, click **Mail** on the menu bar and then click **Settings**. Click **Viewing** (∞) to display the Viewing pane, go to the View Conversations area, and then select (☑) or deselect (☐) **Show most recent message at the top**. Click **Close** (●) to close the Settings window.

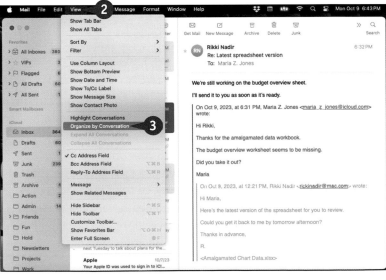

Mail organizes the messages into conversations, so that each exchange appears as a single item rather than as separate messages.

C The number to the right of a conversation indicates how many messages it contains.

4 Click the conversation.

Note: You can expand all conversations in the folder by clicking **View** and clicking **Expand All Conversations**. Click **View** and click **Collapse All Conversations** to collapse them again.

D All the messages in the conversation appear in summary.

5 Click the number to the right of the conversation.

E Mail expands the conversation so that you can see each of the messages it contains.

F You can click a message to display it in the reading pane.

Note: You can mute a conversation by Control +clicking it and then clicking **Mute** on the contextual menu.

Are there other advantages to viewing an exchange as a conversation?
When you view an exchange as a conversation, you can manipulate all the messages in a single move instead of having to manipulate each message individually. For example, click the conversation and drag it to a folder to file all its messages in that folder, or click the conversation and press ⌘ + Delete to delete the entire conversation.

Block and Unblock Contacts

If you receive unwanted e-mail messages from a contact, you may want to block that contact. You can choose whether the blocking simply marks the message as blocked, with the message still appearing in your inbox, or whether Mail puts the message in your Trash folder.

You can unblock a contact quickly either by starting from a message from that contact or by opening the Junk Mail pane in the Settings window and working on the Blocked tab.

Block and Unblock Contacts

Block a Contact

1 In Mail, click the message.

The message appears in the reader pane.

2 Press Control +click the contact's name.

The pop-up menu opens.

3 Click **Block Contact**.

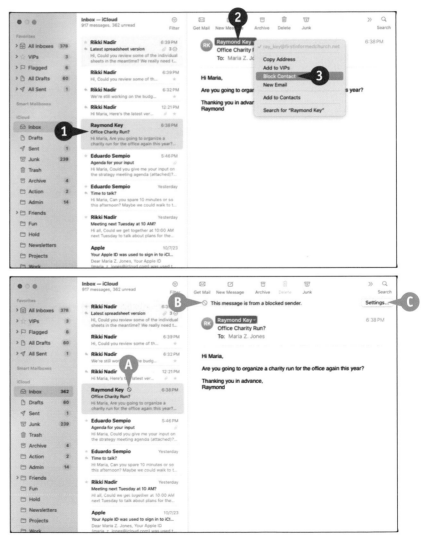

Ⓐ The Blocked icon (🚫) appears on the message in the preview pane.

Ⓑ A blue bar saying *This message is from a blocked sender* appears across the top of the reader pane.

Ⓒ You can click **Settings** to display the Blocked tab in the Junk Mail pane in the Settings window.

Unblock One or More Contacts

1 In Mail, click a message from the blocked contact.

The message appears in the reader pane.

2 Press Control+click the contact's name.

The pop-up menu opens.

3 Click **Unblock Contact**.

4 To unblock multiple contacts, click **Settings** if the button appears. Otherwise, click **Mail** on the menu bar, click **Settings**, click **Junk Mail** (⊟), and then click **Blocked**.

The Blocked tab of the Junk Mail pane appears.

5 Click the contact or contacts you want to unblock.

6 Click **Remove** (—).

Mail removes the contact from the block list.

D In the When Email from Blocked Address Arrives area, click **Mark as blocked mail, but leave it in my Inbox** (◉) or **Move it to the Trash** (◉), as appropriate.

7 Click **Close** (●).

The Settings window closes.

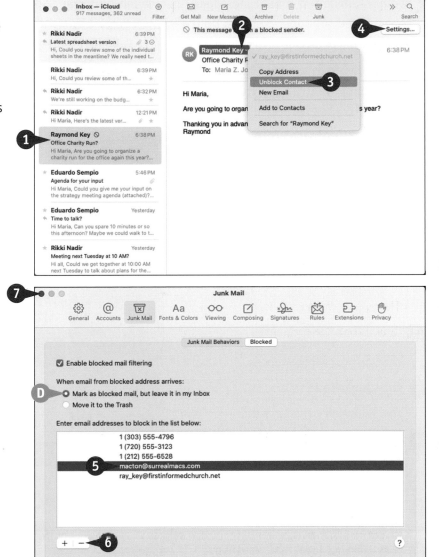

TIP

Why does the Block Contact command not appear for some of my contacts?
The Block Contact command does not appear if the contact is a VIP. Press Control+click the contact's name in a message header to display the pop-up menu, and then click **Remove from VIPs**. You can then press Control+click the name again and click **Block Contact** to block the now-demoted contact.

Reduce the Amount of Spam You Receive

Spam is unwanted e-mail messages, also called *junk mail*. Spam ranges from messages offering specialized products, such as pharmaceuticals, to attempts to steal your financial details, passwords, or personal information.

Mail includes features that enable you to reduce the amount of spam that reaches your inbox. You can configure Mail to identify junk mail automatically, and you can learn to spot identifying features of spam messages. Unfortunately, it is not yet possible to avoid spam completely.

Reduce the Amount of Spam You Receive

Set Mail to Identify Junk Mail Automatically

1 With the Mail app active, click **Mail**.

The Mail menu opens.

2 Click **Settings**.

The Settings window opens.

3 Click **Junk Mail** (⊠).

The Junk Mail pane appears.

4 Click **Junk Mail Behaviors**.

The Junk Mail Behaviors tab appears.

5 Select **Enable junk mail filtering** (☑).

6 Click **Mark as junk mail, but leave it in my Inbox** (○ changes to ◉) to review junk mail in your inbox.

7 Select each of the three check boxes (☑).

8 Select **Trust junk mail headers in messages** (☑).

9 Click **Close** (●).

The Settings window closes.

Review Your Junk Mail

1 Click **Inbox** or **Junk** of a particular account, such as **iCloud**.

The messages appear.

2 Click a message.

A The message *Unable to load remote content privately* indicates that the sender is trying to use remote content to track the message.

3 Check whether the message greets you by name, with a generic greeting, or not at all.

4 Read and evaluate the title of the message.

5 Read the message's content for veracity.

Note: Junk mail often contains typos, odd phrasing, unusual or missing punctuation, or unorthodox grammar.

6 If a message appears to be spam, and Mail has not identified it as junk, click **Junk** (🗑). Conversely, if Mail has identified the message as junk, but it is not, click **Not Junk** (🗑).

7 Click **Delete** (🗑) to delete the message.

TIP

How can I tell whether a message is genuine or spam?

If a message does not show your e-mail address and your name, it is most likely spam. If the message does show your e-mail address and name, read the content carefully to establish whether the message is genuine. If the message calls for action, such as reactivating an online account that you have, do not click a link in the message. Instead, open Safari and type the address of the website, making sure it is correct — if in doubt, use a search engine to search for the address. Log in to the website as usual, and see if an alert is waiting for you.

CHAPTER 8

Chatting and Calling

macOS includes Messages for instant messaging and FaceTime for video chat with users of Macs, iPhones, and iPads. You can also configure Handoff with your iPhone so that you can phone and send SMS messages using your MacBook.

Configure the Messages App

The Messages app enables you to chat with your contacts via instant messaging. Using Messages, you can connect via Apple's iMessage service to contacts on Macs, iPhones, and iPads.

If you have set up iCloud on your MacBook, Messages should already be configured for your iCloud account. If you have not set up iCloud, you must add an account to Messages before you can use the app.

Configure the Messages App

1 Click **Messages** (◯) on the Dock.

The Sign In to iMessage with Your Apple ID dialog opens.

Note: If the Messages window opens, go to step **5**.

2 Type your Apple ID.

3 Click **Sign In**.

Ⓐ You can click **Forgot Apple ID or Password?** or **Forgot password?** if you have forgotten your credentials.

Ⓑ The Password field appears.

4 Type your password.

5 Click **Sign In**.

The Messages window opens.

6 Click **Messages** to open the Messages menu.

7 Click **Settings** to display the Settings window.

8 Click **General** (⚙) to show the General tab.

9 Click **Set up Name and Photo Sharing**, and then follow the prompts to set your contact details and photo, icon, or memoji.

10 Click **Keep messages** (◆) and click the length of time to keep messages: **30 Days**, **One Year**, or **Forever**.

11 Select (☑) or deselect (☐) **Notify me about messages from unknown contacts**, as appropriate.

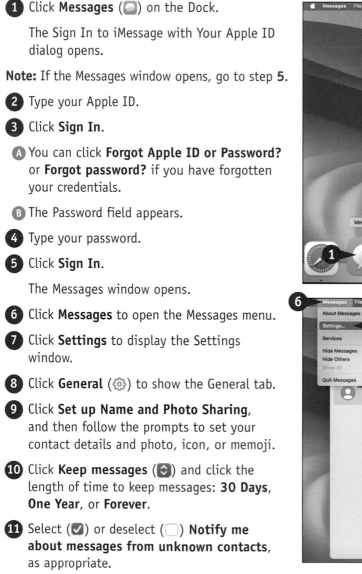

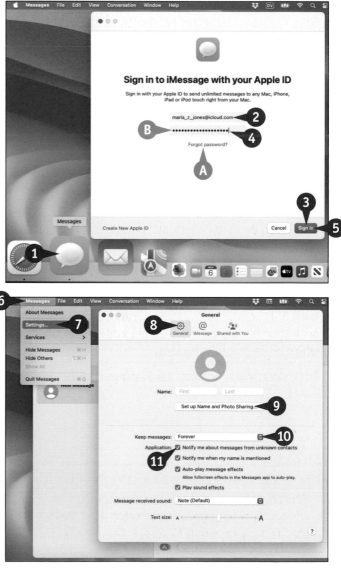

224

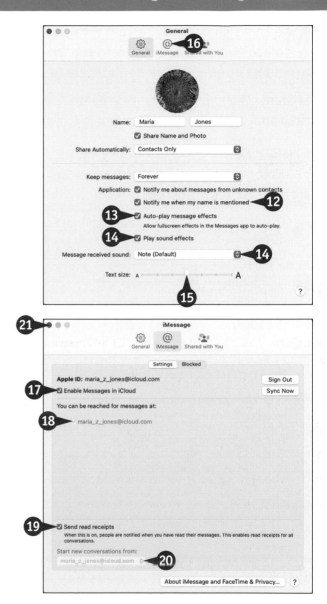

12 Select **Notify me when my name is mentioned** (☑) to have Messages tell you when your first name or full name appears in a message. This can be useful in group chats.

13 Select **Auto-play message effects** (☑) to allow full-screen effects to play automatically.

14 Select (☑) **Play sound effects** if you want to hear sound effects. Click **Message received sound** (◉) and select the sound for incoming messages.

15 Drag the **Text size** slider as needed to set the text size for Messages.

16 Click **iMessage** (@) to display the iMessage tab.

17 Select **Enable Messages in iCloud** (☑) if you want to sync your messages via iCloud so that all your devices can show the same conversations.

18 In the You Can Be Reached for Messages At list, deselect (☐) any phone number or address at which you do not want to be contacted.

19 Select **Send read receipts** (☑) if you want to send read receipts for messages in all conversations.

20 Click **Start new conversations from** (◉) and then click the phone number or e-mail address from which to start conversations.

21 Click **Close** (●) to close the Settings window.

TIP

What settings can I configure on the Shared with You tab of the Settings dialog?

The Shared with You tab lets you choose which apps can automatically receive content shared with you in the Messages app. For example, if you select (☑) **Photos** on this tab, the Photos app can automatically receive images shared with you in Messages. On this tab, select (☑) or deselect (☐) the check box for each app, as needed. Alternatively, click **Turn Off** to turn off automatic sharing for all apps.

Chat with a Contact

Messages enables you to chat with your contacts via instant messaging. The easiest way to start using Messages is by sending text messages. Depending on the messaging services and the computers or devices your contacts are using, you may be able to chat via audio or video as well.

To start chatting, you send your contact an invitation. If your contact accepts the invitation, the reply appears in the Messages window. You can conduct multiple chats simultaneously, switching from chat to chat as needed.

Chat with a Contact

1 Click **Messages** (💬) on the Dock.

The Messages window opens.

2 Click **Compose New Message** (📝).

A New Message entry appears in the left pane.

3 Click **Add Contact** (⊕).

The Contacts panel opens.

4 Click the contact with whom you want to chat.

The contact's addresses appear.

5 Click the appropriate address.

A The contact's name appears in the To area.

Note: If the contact's name appears in red, it means that the contact is not registered with iMessage. You may need to use a different address to reach the contact.

6 Type the text you want to send, and then press Return.

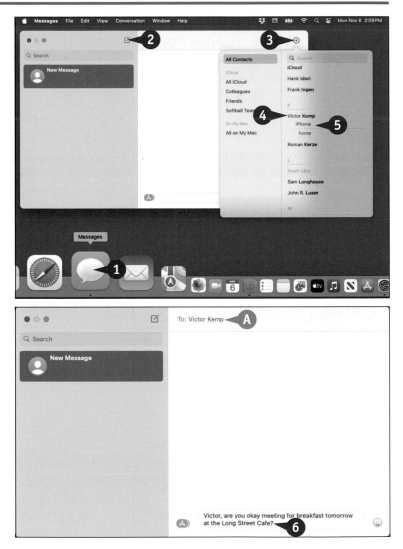

B Your message appears in a bubble on the right side of the right pane.

C A reply from your contact appears on the left side of the right pane.

7 Type a reply to your contact's reply.

8 If you would like to add a special character, click **Special Characters** (☺).

The Special Characters panel opens.

9 Click the category of special characters you want: **Recents and Favorites** (🕐), **Smileys & People** (☺), **Animals & Nature** (🐻), **Food & Drink** (🍔), **Activity** (⚽), **Travel & Places** (🏛), **Objects** (💡), **Symbols** (🔣), or **Flags** (🏳).

10 Click the special character you want to use.

Note: Press and hold a special character to display a pop-up panel containing any alternative forms it offers. For example, some special characters offer a selection of skin tones.

The special character appears in the message.

11 Press Return to send the message (not shown).

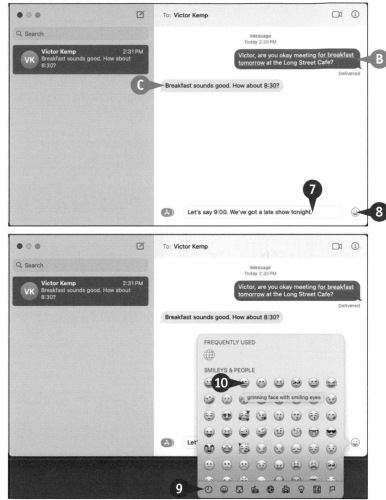

Send and Receive Files with Messages

As well as chat, Messages enables you to send files easily to your contacts and receive files they send to you. During a chat, you can send a file either by using the Send File command or by dragging a file from a Finder window into Messages.

When a contact sends you a file, you can decide whether to receive it. Messages automatically stores the files you receive in the Downloads folder in your user account, but you can change the destination to another folder if you so choose.

Send and Receive Files with Messages

Send a File

1 Start a text chat with the contact to whom you want to send a file, or accept a chat invitation from that contact.

2 Click **Conversation**.

The Conversation menu opens.

3 Click **Send File**.

The Send File dialog opens.

4 Click the file you want to send.

5 Click **Open**.

A A button for the file appears in the text box.

6 Type any message needed.

7 Press **Return** (not shown).

Messages sends the message, including a button for transferring the file.

If your contact accepts the file, Messages transfers it.

Receive a File

B When your contact sends you a file, it appears as a button in the Chat window.

1 Press **Control**+click the file's button.

The contextual menu opens.

2 Click **Save to Downloads**.

Messages saves the file to the Downloads folder.

You can then open the Downloads folder — for example, click **Finder** (🙂) on the Dock, and then click **Downloads** in the sidebar — and double-click the file to open it in its default app.

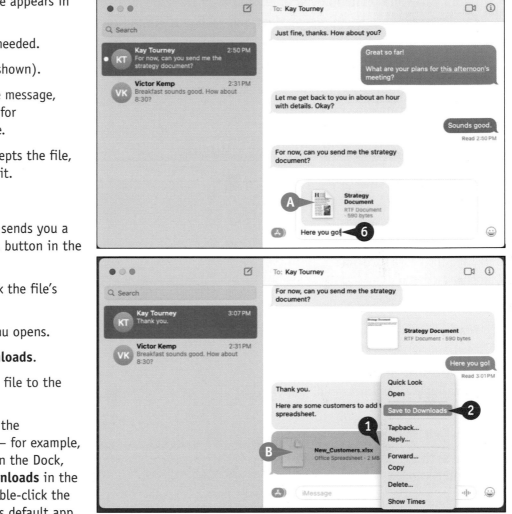

TIP

How do I block a contact in Messages?

In the left pane, click a conversation from the contact so that the conversation's contents appear in the right pane. Click **Conversation** on the menu bar to open the Conversation menu, and then click **Block Person**. In the Are You Sure You Want to Block *Name*? dialog, click **Block**.

To unblock the contact, click a conversation from them in the left pane, click **Conversation** on the menu bar, and then click **Block Person**, removing the check mark. Alternatively, click **Messages** on the menu bar, click **Settings** to open the Settings window, and then click **iMessage** (@). Click **Blocked** to display the Blocked tab, click the contact, and then click **Remove** (—).

Sign In to FaceTime and Choose Settings

Apple's FaceTime technology enables you to make audio and video calls easily across the Internet. FaceTime works with all recent Macs and with current and recent models of the iPhone and iPad. The Live Captions feature can display a text readout of the conversation on screen in almost real time.

Your MacBook includes a built-in video camera and microphone, so it is ready to use FaceTime right out of the box. Before you can make calls, you may need to sign in to FaceTime. You may also want to configure settings for FaceTime.

Sign In to FaceTime and Choose Settings

Open and Set Up FaceTime

1 Click **FaceTime** (⬛) on the Dock.

Note: If the FaceTime icon does not appear on the Dock, click **Launchpad** (⬛) on the Dock, and then click **FaceTime** (⬛) on the Launchpad screen.

    The FaceTime window opens.

    The FaceTime app signs you in to the FaceTime service.

Note: If FaceTime prompts you to sign in, type your Apple ID and password, and then click **Sign In**.

2 Click **FaceTime** on the menu bar.

3 Click **Settings** to open the FaceTime Settings window.

4 Click **General** to display the General tab.

5 Make sure **Enable this account** is selected (☑).

6 In the You Can Be Reached for FaceTime At list, select (☑) or deselect (☐) each phone number or Apple ID, as needed.

7 In the Automatic Prominence section, select **Speaking** (☑) if you want FaceTime to enlarge the tile of the person speaking in a group call.

8 Select (☑) or deselect (☐) **Live Captions**, as needed.

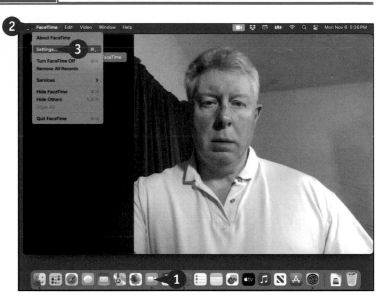

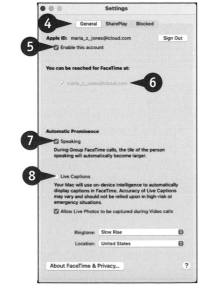

9 Select (☑) or deselect (☐) **Allow Live Photos to be captured during Video calls** to control whether participants in video calls can capture photos during calls. See the tip for more information.

10 Click **Ringtone** (◉) and choose the ringtone you want to use.

11 Verify that the **Location** pop-up menu (◉) shows your location. If not, click **Location** (◉) and click your location.

12 Click **SharePlay**.

The SharePlay tab appears.

13 Select **SharePlay** (☑) if you want to use FaceTime to share shows, music, and apps with others.

14 In the list box, adjust the list of apps, as needed.

15 Click **Blocked**.

The Blocked tab appears.

Ⓐ To remove a phone number or an Apple ID from the blocked list, click the appropriate line and then click **Remove** (—).

Ⓑ To block a phone number or an Apple ID, click **Add** (+), and then click the contact you want to block. The contact's addresses appear in the list.

16 Click **Close** (⬤).

The Settings window closes.

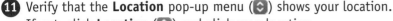

TIP

What does the Allow Live Photos to Be Captured During Video Calls setting do?
Unlike "live" photos in the Photos app, which include several seconds of video around a still image captured on an iPhone or iPad, "live" photos in FaceTime refers to taking still shots by tapping **Take Photo** (◉) during a FaceTime video call. If you deselect (☐) this check box, no participant in the call can take still photos in the FaceTime call. Instead, the Live Photos Is Disabled message box appears, prompting you to tap **Enable** to enable Live Photos for future calls. Any participant can still take screenshots of the call — for example, by pressing ⌘+Shift+3 to capture the full screen on a Mac.

Make and Receive FaceTime Calls

When you have set up FaceTime with your Apple ID, you can make and receive FaceTime calls from your MacBook. You can call any iPhone user or any user of a Mac or iPad who has enabled FaceTime.

To make a call, you open FaceTime, click the contact, and then select the e-mail address or phone number to use for contacting them. If the contact uses Windows, you create a link and send it via e-mail or messaging. To receive a call, you simply answer when FaceTime alerts you to the incoming call.

Make and Receive FaceTime Calls

Make a FaceTime Call

1 Click **FaceTime** (▢) on the Dock.

Note: If no FaceTime icon appears on the Dock, click **Launchpad** (▦), and then click **FaceTime** (▢).

2 Click **New FaceTime** (▭).

Ⓐ You can click **Create Link** (🔗) to create a link to send to a contact who uses Windows.

The New FaceTime dialog opens.

3 Start typing the contact's name, and then click the contact.

4 Click **FaceTime**.

Ⓑ You can make an audio call by clicking ▾, and then clicking **FaceTime Audio**.

FaceTime places the call.

5 When your contact answers, begin chatting (not shown).

C Your video preview appears as a thumbnail.

D You can tap **Take Photo** (◯) to take a photo of the chat.

6 When you are ready to finish the call, move the mouse to display the pop-up controls, and then click **End** (⊗).

Receive a FaceTime Call

When you receive a FaceTime call, a FaceTime window opens.

1 Click **Accept**.

Note: In the Contacts app, you can assign a specific ringtone and a specific text tone to a contact. These tones can help you identify which contact is calling or messaging you.

E If you want an audio-only call, you can click ⬇, and then click **Answer as Audio**.

F If you want to decline the call, click **Decline**.

G If you want to postpone the call, click ⬇, and then click **Remind Me in 5 Minutes**, **Remind Me in 15 Minutes**, or **Remind Me in 1 Hour**.

H If you want to text the caller, click ⬇, and then click **Reply with Message**.

TIPS

Must I keep FaceTime running all the time to receive incoming calls?
No. After you set up FaceTime with your Apple ID, FaceTime runs in the background even when the app itself is not open. When FaceTime detects an incoming call, it plays a ringtone and displays a notification window in front of your other open windows. You can then decide whether to accept the call or reject it.

How do I stop using FaceTime temporarily on one of my Macs?
You can turn off FaceTime temporarily by clicking **FaceTime** (◻) on the Dock to launch FaceTime, clicking **FaceTime** on the menu bar, and then clicking **Turn FaceTime Off**. When you want to resume using FaceTime, click **FaceTime** on the menu bar and click **Turn FaceTime On**.

Configure and Use Handoff with Your iPhone

If you have an iPhone or iPad, you can enjoy the impressive integration that Apple has built into the iOS, iPadOS, and macOS operating systems. Apple calls this integration Continuity. Continuity involves several features including Handoff, which enables you to pick up your work or play seamlessly on one device exactly where you have left it on another device. For example, you can start writing an e-mail message on your Mac, continue it on your iPhone, and then complete it on your iPad.

Understanding Which iPhone, iPad, and Mac Models Can Use Continuity

All current and recent iPhone models, iPad models, and Mac models can use Continuity. Your Mac must have Bluetooth 4.0 hardware. If your Mac can run macOS Monterey or a later version, such as macOS Ventura or macOS Sonoma, it can use Continuity.

Enable Handoff on Your iPhone

To enable your iPhone to communicate with your Mac, you need to enable the Handoff feature. From the Home screen, tap **Settings** (⚙) to open the Settings app, tap **General** (⚙) to display the General screen, and then tap **AirPlay & Handoff** (A). On the AirPlay & Handoff screen, set the **Handoff** switch (B) to On (⬤).

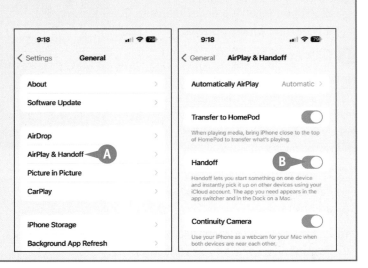

Enable Handoff on Your Mac

You also need to enable Handoff on your Mac. To do so, click **Apple** () on the menu bar and then click **System Settings** to open the System Settings window. Click **General** (C, ⚙) to display the General pane, and then click **AirDrop & Handoff** to display the AirDrop & Handoff pane. Set the **Allow Handoff between this Mac and your iCloud devices** switch (D) to On (⬤). You can then click **Close** (E, ⬤) to close System Settings.

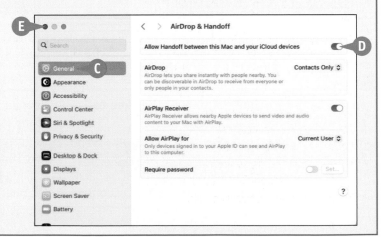

Make and Take Phone Calls on Your Mac

When you are using your Mac within Bluetooth range of your iPhone, Continuity enables you to make and take phone calls on your Mac instead of your iPhone. For example, when someone calls you on your iPhone, your Mac displays a call window automatically, and you can pick up the call on your Mac.

Send and Receive Text Messages from Your Mac

Your Mac can already send and receive messages via Apple's iMessage service, but when your iPhone's connection is available, your Mac can send and receive messages directly via Short Message Service (SMS) and Multimedia Messaging Service (MMS). This capability enables you to manage your messaging smoothly and tightly from your Mac.

Your SMS conversations appear in green bubbles, whereas iMessage conversations appear in blue bubbles, enabling you to distinguish SMS conversations easily from iMessage conversations and to know that you cannot use iMessage features, such as file exchange, in SMS conversations.

CHAPTER 9

Organizing Your Life

To help you keep your daily life organized, your MacBook includes the Calendar, Contacts, Reminders, and Maps apps.

Manage Your Apple ID and iCloud Account

Your Apple ID is your personal account for accessing Apple's services, from iCloud to the iTunes store and from FaceTime to iMessage. Your MacBook uses your Apple ID for authentication, authorization, and payments on Apple's services, so you should make sure that your Apple ID information is accurate. You can do so by working in the Apple ID pane in the System Settings app.

Display the Apple ID Pane in System Settings

Click **System Settings** (A, ⚙) on the Dock to open the System Settings window, and then click **Apple ID** (B) in the sidebar to display the Apple ID pane.

Your picture (C), emoji, or monogram appears at the top of the pane, with your name and the e-mail address used for your Apple ID (D) below it. Below the e-mail address are three buttons for navigating the categories of Apple ID settings: Personal Information (E, 🪪), Sign-In & Security (F, 🔘), and Payment & Shipping (G, 🖼).

In the next section of the pane are buttons for iCloud (H, ☁), Media & Purchases (I, 🅰), and Family Sharing (J, 👥). Below this section is the Devices list (K), which shows a button for each device associated with your Apple ID. Devices include Macs, PCs, iPhones, iPads, and Apple Watches.

Configure Personal Information Settings and Communication Preferences

Click **Personal Information** (🪪) to display the Personal Information pane. To change your memoji, photo, or monogram, click **Edit** (L) and work in the dialog that opens. To change your birthday, use the date picker (M).

To choose which announcements, news updates, and promotional information you receive from Apple's services, click **Details** (N, ⓘ) in the Communication Preferences box. In the Communication Preferences dialog, set the **Announcements** switch (O), the **Apps, Music, TV, and More** switch (P), and the **Apple News Newsletter** switch (Q) to On (🔘) or Off (⚪), as needed, and then click **Done** (R) to close the dialog.

Click **Back** (S, ‹) to return to the Apple ID pane.

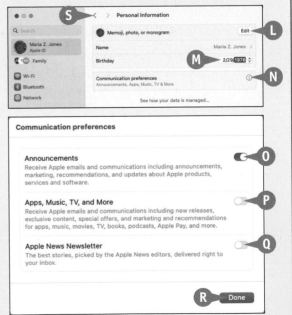

Configure Sign-In & Security Settings

Click **Sign-In & Security** (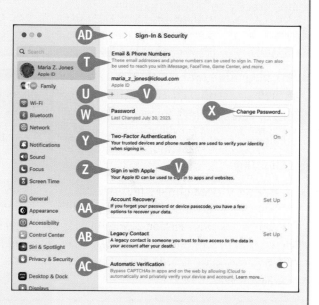) in the Apple ID pane to display the Sign-In & Security pane. Here, the Email & Phone Numbers box (T) shows the e-mail addresses and phone numbers you can use to sign in to your account. You can click **Add** (U, +) to add a new e-mail address or phone number, or click **Remove** (V, —) to remove the selected address or number.

The Password box (W) shows the date you last changed your password. You can click **Change Password** (X) to change it now.

Verify that the Two-Factor Authentication readout (Y) shows On to make sure that your trusted devices — such as your iPhone and iPad — are being used to verify your identity when you sign in on untrusted devices. If not, click this button and follow the prompts to enable Two-Factor Authentication.

The Sign in with Apple box (Z) shows whether you have set up your Apple ID to sign in to apps and websites.

The Account Recovery box (AA) shows whether you have set up a means to recover your Apple ID and data if you lose access to your account. If Set Up appears, click this button and use the Account Recovery dialog to specify one or more Recovery Contacts, create a 28-character Recovery Key code, or both.

The Legacy Contact box (AB) shows whether you have specified a legacy contact, someone you trust to access and administer your data after your death. If Set Up appears, click this button, and then use the Legacy Contact dialog to add a contact.

The Automatic Verification switch (AC) shows whether you are allowing iCloud to automatically and privately verify your device and account to bypass CAPTCHA prove-you-are-human-not-a-bot tests. Set this switch to On (⬤) or Off (⭘), as needed.

Click **Back** (AD, <) to return to the Apple ID pane.

Configure Payment & Shipping Settings

Click **Payment & Shipping** (▭) to display the Payment & Shipping pane. Verify that your payment method, Family Sharing payment method, and shipping address are correct.

Navigate the Calendar App

The Calendar app enables you to input your appointments and events and track them easily. After launching the app, you can navigate to the dates you need to work with. You can sync your calendar data with your iPhone or iPad.

Calendar has a streamlined user interface that makes it easy to move among days, weeks, months, and years. You can click the **Today** button to display the current day, or use the Go to Date dialog to jump directly to a specific date.

Navigate the Calendar App

Open Calendar and Navigate by Days

1 Click **Calendar** (📅) on the Dock.

Calendar opens.

2 Click **Day**.

Calendar displays the current day, including a schedule of the day's events.

3 Click **Next** (>) to display the next day or **Previous** (<) to display the previous day.

View and Navigate by Weeks

1 Click **Week**.

Calendar displays the week for the date you were previously viewing.

2 Click **Next** (>) to display the next week or **Previous** (<) to display the previous week.

Ⓐ The Time Zone pop-up menu (↕) enables you to switch between time zones. To display this pop-up menu, click **Calendar** on the menu bar, click **Settings**, click **Advanced** (⚙) and then select **Turn on time zone support** (☑).

Ⓑ An event after the current times displayed appears at the bottom of the window. Scroll down to view it.

View and Navigate by Months

1 Click **Month**.

Calendar displays the current month.

2 Click **Next** (>) to display the next month or **Previous** (<) to display the previous month.

View and Navigate by Years

1 Click **Year**.

Calendar displays the year for the date you were last viewing.

2 Click **Next** (>) to display the next year or **Previous** (<) display the previous year.

Note: In Year View, the dates shown in bold include events. The dates shown in gray have no events yet.

Note: In Week view, Month view, or Year view, double-click a day to display it in Day view.

C The red circle shows today's date.

D You can click **Today** to display today's date.

TIP

Which keyboard shortcuts can I use to navigate in Calendar?
Press ⌘+1 to display the calendar by day, ⌘+2 by week, ⌘+3 by month, or ⌘+4 by year. Press ⌘+→ to move to the next day, week, month, or year, or press ⌘+← to move to the previous one. Press ⌘+Shift+T to open the Go to Date dialog, which enables you to jump to a specific date. Press ⌘+T to jump to today's date.

Create a New Calendar

Calendar enables you to create as many calendars as you need to separate your events into logical categories. Calendar comes with three iCloud calendars already created for you: the Home calendar, the Family calendar, and the Work calendar. Any calendars in other online accounts you have set up on your MacBook and enabled calendars on appear automatically as well. You can create new calendars as needed alongside these calendars.

After creating a new calendar, you can create events in it. You can also change existing events from another calendar to the new calendar.

Create a New Calendar

1 Click **Calendar** (📅) on the Dock.

Calendar opens.

2 Click **File** on the menu bar.

The File menu opens.

3 Click **New Calendar**.

Note: If, when you click **New Calendar**, the New Calendar continuation menu opens, click the calendar service, such as iCloud, in which to create the new calendar.

Note: If the Calendars pane is open, you can create a new calendar by pressing `Control`+clicking in open space in the Calendars pane and then clicking **New Calendar** on the contextual menu.

Ⓐ Calendar displays the Calendars pane if it was hidden, as in this example.

Calendar creates a new calendar and displays an edit box around its default name, Untitled.

4 Type the name for the calendar and press `Return`.

Calendar applies the name to the calendar.

Ⓑ At any time, you can click **Calendars** (📅) to toggle the display of the Calendars pane.

5 Press `Control`+click the calendar's name.

The contextual menu opens.

6 Click **Get Info**.

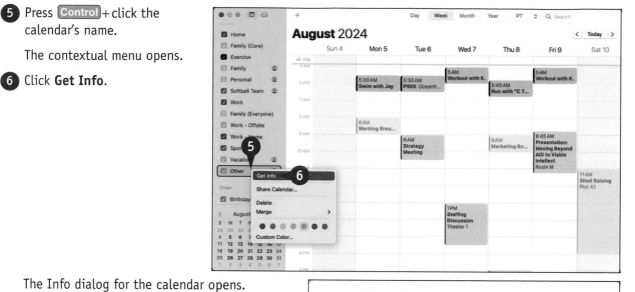

The Info dialog for the calendar opens.

7 Click **Color** (⚬) and then select the color you want the calendar to use.

8 Type a description for the calendar.

Note: July 17, 2002 was the date on which Apple first announced iCal, its calendar app that later became the Calendar app. July 17 is now informally celebrated as World Emoji Day.

C Select **Ignore alerts** (☑) if you want to suppress alerts for the calendar.

9 Click **OK**.

The dialog closes.

You can now add events to the calendar.

TIPS

What do the check boxes in the Calendars pane do?

The check boxes control what calendars Calendar displays. Deselect a check box (☐) to remove a calendar's events from display.

What does the Merge command on the pop-up menu for a calendar do?

The Merge command enables you to merge the contents of one calendar into another calendar. For example, if you have two calendars whose content areas overlap, you might merge them into a single calendar. After merging the calendars, the Calendar app deletes the calendar from which you start the merge operation.

Create an Event

Calendar makes it easy to organize your time commitments by creating an event for each appointment, meeting, trip, or special occasion. Calendar displays each event as an item on its grid, so you can see what is supposed to happen when.

You can create an event either for a specific length of time, such as 1 hour or 2 hours, or for an entire day. And you can create either an event that occurs only once or an event that repeats one or more times, as needed.

Create an Event

1 Click **Calendar** (📅) on the Dock.

Calendar opens.

2 Navigate by days, weeks, months, or years to reach the day on which you want to create the event.

3 Click **Day**.

Calendar switches to Day view.

4 Click the event's start time and drag to its end time.

Calendar creates an event where you clicked and applies a default name, New Event.

When you release the trackpad button, the event's details appear in the right pane.

5 Type the name for the event and then press Return.

6 Click **Add Location or Video Call** and enter the location.

7 Click **Calendar** (↕) and then click the calendar to which you want to assign the event.

8 Click **Add Alert, Repeat, or Travel Time**.

Controls for setting the alert, repeat, and travel time appear.

Ⓐ You can select **all-day** (☑) to make the event an all-day event.

Ⓑ You can click **travel time** and specify the travel time required.

⑨ If you want a reminder, click **alert** and specify the details of the alert, such as **10 minutes before**. Choose **Time to Leave** to have macOS calculate the travel time to the address you have specified.

⑩ Click **Add Invitees** and specify anybody to invite to the event.

⑪ To add more information, click **Add Notes, URL, or Attachments**.

⑫ Click **Add Note** and type any notes needed.

⑬ Click **Add URL** and type or paste the URL for the event.

⑭ Click **Add Attachment**, click the file in the Open dialog, and then click **Open**.

⑮ When you finish entering details, click outside the details pane.

The details pane closes.

How do I create a repeating event?

In the details pane, click **repeat** and click **Every Day**, **Every Week**, **Every Month**, or **Every Year**. Use the controls that appear for setting the details of the repetition — for example, click **end repeat**, click **After**, and then specify **8 times**. For other options, click **Custom** to open the Custom dialog. You can then click **Frequency** (⬍) and select **Daily**, **Weekly**, **Monthly**, or **Yearly**, and then specify the repetition patterns, such as **Every 2 Weeks** or **Every 1 month(s) On the first Monday**.

Share an iCloud Calendar with Other People

Calendar enables you to share any calendar stored on iCloud with other people so that they know when you are busy. You can share an iCloud calendar either as a private calendar, available only to the people whose names or e-mail addresses you specify, or as a public calendar, available to everyone.

Share an iCloud Calendar with Other People

Open the Dialog for Sharing a Calendar

1 Click **Calendar** (📅) on the Dock.

Calendar opens.

2 Click **Calendars** (📖).

The Calendars pane opens.

3 Press **Control** +click the calendar you want to share.

The contextual menu opens.

4 Click **Share Calendar**.

The Share dialog opens.

Share a Calendar with Specific People

1 In the Share dialog, start typing a name or e-mail address.

2 Click the e-mail address for the contact.

The contact's name appears as a button in the Shared With list.

3 Click the pop-up button (⌄) on the contact's button.

The pop-up menu opens.

4 Click **View & Edit** to enable the contact to edit the calendar. Click **View Only** to enable the contact to only view the calendar.

Ⓐ You can add other contacts as needed by clicking **Share With** (⊕).

5 Click **Done**.

Calendar shares the calendar with the people you specified.

Make a Calendar Public

1 In the Share dialog, select **Public Calendar** (☑).

2 Click **Done**.

The Share dialog closes.

Calendar makes the calendar public.

3 Click **Sharing** (●).

The Share dialog opens again.

Ⓑ The URL field appears, showing the web address for the shared calendar.

Ⓒ You can click **Share** (⬆) to share the URL.

4 Click **Done**.

The Share dialog closes again.

TIPS

How do I tell people about a calendar I have made public?
Click **Calendars** (▦) to display the Calendars pane. Press `Control`+click the calendar and click **Send Publish Email**. Your default e-mail app creates a new message titled *Subscribe to my online calendar* that contains a link to the calendar. Address the message; if you are sending a group message, add yourself as the To recipient and put all other recipients in the Bcc field so each sees only their own e-mail address. Add any other information needed, and then send the message.

How do I stop a public calendar from being public?
In the Calendars pane, `Control`+click the calendar, click **Unpublish**, and then click **Stop Publishing** in the confirmation message box that opens.

Subscribe to a Shared Calendar

Calendar enables you to subscribe to calendars that others have shared on iCloud or published on the Internet. By subscribing to a calendar, you add it to Calendar so that you can view the events in the calendar along with those in your calendars.

You can subscribe to a calendar either by typing or pasting its URL into Calendar or by clicking a link in a message that you have received.

Subscribe to a Shared Calendar

1 Click **Calendar** (📅) on the Dock.

Calendar opens.

2 Click **File**.

The File menu opens.

3 Click **New Calendar Subscription**.

Note: Many organizations, sports teams, and artists make their calendars available on their websites. You can either copy the calendar's URL or download the calendar.

The Enter the URL of the Calendar You Want to Subscribe To dialog opens.

4 Paste or type in the calendar's URL.

Note: If you receive a link to a published calendar, click the link in Mail. Calendar opens and displays the Enter the URL of the Calendar You Want to Subscribe To dialog with the URL inserted. Click **Subscribe**.

5 Click **Subscribe**.

A dialog opens showing the details of the calendar.

6 If necessary, edit the default name to display for the calendar.

7 Click **Color** (◉) and select the color to use for the calendar.

8 Click **Location** (◉) and choose where to store the calendar. Your choices are **iCloud** or **On My Mac**.

9 Select **Alerts** (☑) if you want to remove alerts.

10 Select **Attachments** (☑) if you want to remove attachments.

11 Click **Auto-refresh** (◉) and select your preferred option for automatically refreshing the calendar, such as **Every day** or **Every week**.

12 Select **Ignore alerts** (☑) if you want to ignore alerts set in the calendar.

13 Click **OK**.

Calendar adds the calendar, and its events appear.

"MLB Playoffs" Info

Name: MLB Playoffs — **6** — **7**

Subscribed to: https://calendar.google.com/calendar/ical/5c7c

Location: iCloud — **8**

Remove: ☑ Alerts — **9**
☑ Attachments

Last updated: Never

Auto-refresh: Every week

☐ Ignore alerts

Cancel   OK

"MLB Playoffs" Info

Name: MLB Playoffs

Subscribed to: https://calendar.google.com/calendar/ical/5c7c

Location: iCloud

Remove: ☑ Alerts
☑ Attachments — **10**

Last updated: Never

Auto-refresh: Every day — **11**

12 ☑ Ignore alerts

Cancel   OK — **13**

TIPS

How do I update a published calendar?
You can update all your published calendars at the same time by clicking **View** on the menu bar and then clicking **Refresh Calendars**. You can also give the Refresh Calendars command by pressing ⌘+R.

How do I unsubscribe from a published calendar?
Click **Calendars** to display the Calendars pane, press Control +click the calendar, and then click **Unsubscribe** on the contextual menu.

Add Someone to Your Contacts

The Contacts app enables you to track and manage your contacts. Contacts stores the data for each contact on a separate virtual address card that contains storage slots for many different items of information, from the person's name and phone numbers to the e-mail addresses and photo.

To add a contact, you create a new contact card and enter the person's data on it. You can also add contact information quickly from vCard address card files that you receive.

Add Someone to Your Contacts

 Click **Contacts** (⬛) on the Dock to open the Contacts app.

② Click **Add** (**+**) to open the pop-up menu.

 Click **New Contact**.

Note: You can also create a new contact card by clicking **File** on the menu bar and then clicking **New Card** on the File menu.

Contacts creates a new card and selects the First placeholder.

④ Type the contact's first name.

Note: Press Tab to move the focus from the current field to the next.

⑤ Type the contact's last name.

⑥ If the contact works for a company, type the company name.

Ⓐ You can select **Company** (☑) when creating a card for a company or organization rather than for an individual. Contacts then uses the company name for sorting.

⑦ Click the pop-up menu (○) next to the first Phone field and select the type of phone number, such as **work** or **mobile**.

⑧ Type the phone number.

Note: You can add other phone numbers as needed.

 Click the picture placeholder.

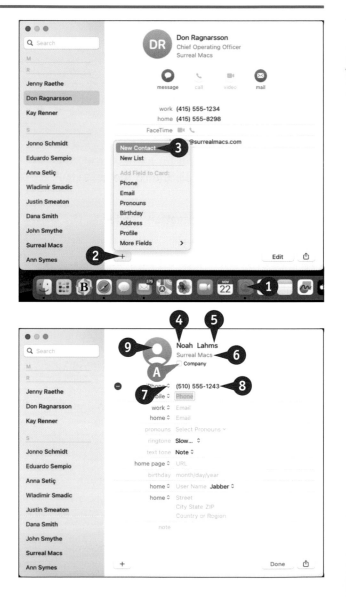

250

The Picture dialog opens.

⑩ In the sidebar, click **Memoji** (😀), **Emoji** (😊), **Monogram** (✏), **Camera** (📷), **Photos** (🖼), or **Suggestions** (🗀), as appropriate.

Note: Click **Camera** (📷) to take a photo with your MacBook's camera. Click **Photos** (🖼) to use a picture from the Photos app.

This example uses **Suggestions** (🗀).

⑪ Click the image to use.

⑫ Drag the slider to zoom in or out.

⑬ Drag the photo to change the part that appears.

⑭ Click **Save**.

The Picture dialog closes.

Ⓑ Contacts adds the picture to the contact record.

⑮ Add any other information needed to the contact record.

Ⓒ If you need a standard field that does not currently appear, click **Add** (➕), click **More Fields**, and then click the field on the continuation menu.

⑯ Click **Done**.

Contacts adds the contact record to your Contacts list.

Note: Only the fields that contain data appear in the card.

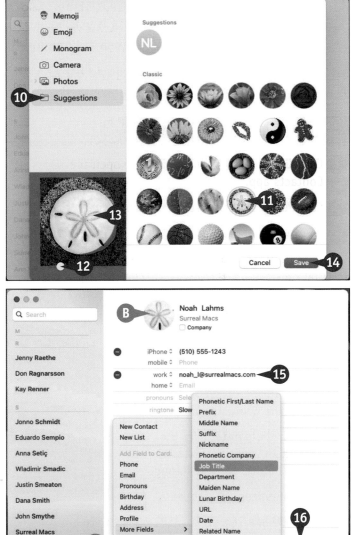

TIPS

How do I add a vCard to Contacts?

If you receive a vCard file, which has the .vcf file extension and contains a virtual address card, in Mail, press `Control`+click the card, highlight **Open With** on the contextual menu, and then select **Contacts**. A dialog opens, prompting you to confirm that you want to import the card into Contacts. Click **Import**.

How do I delete a contact from Contacts?

To delete a contact, click the card, click **Edit**, and then click **Delete Card** or press `Delete`. A confirmation dialog opens. Click **Delete**.

Organize Contacts into Lists

Contacts enables you to organize your contacts into separate lists, making it easier to find the contacts you need. Lists are useful if you have several different categories of contacts, such as family, friends, and colleagues. You can assign any contact to as many lists as needed.

After creating lists, you can view a single list at a time or search within a list. You can also send an e-mail message to all the members of a list.

Organize Contacts into Lists

Create a List of Contacts

1 Click **Contacts** (■) on the Dock.

The Contacts app opens.

2 Click **View**.

The View menu opens.

3 Click **Show Lists**.

Note: You can also press ⌘+Shift+1 to display or hide the Lists pane.

The Lists pane opens on the left side of the Contacts window.

4 Position the pointer over the account in which you want to create the list, such as **iCloud**.

The Add button (➕) appears.

5 Click **Add** (➕).

Contacts adds a list and displays an edit box around the default name, Untitled List.

6 Type the name and press Return.

The name appears.

Note: Your contact lists appear on any iOS device you sync with the same iCloud account. iOS devices enable you to create and edit contacts, but not to manipulate lists.

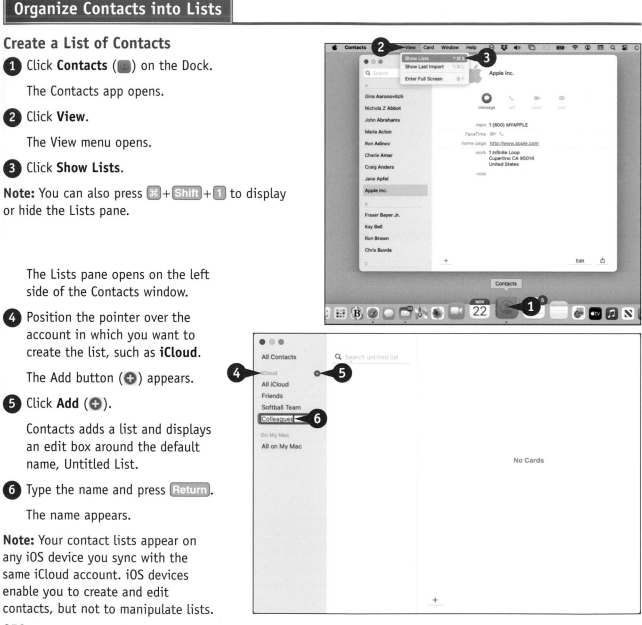

Add Contacts to a List

1 Click **All Contacts**.

Contacts displays all your contacts.

2 Drag one or more contacts to the new list.

Note: To add multiple contacts to the list, click the first, and then press ⌘+click each of the others. Drag the selected contacts to the list.

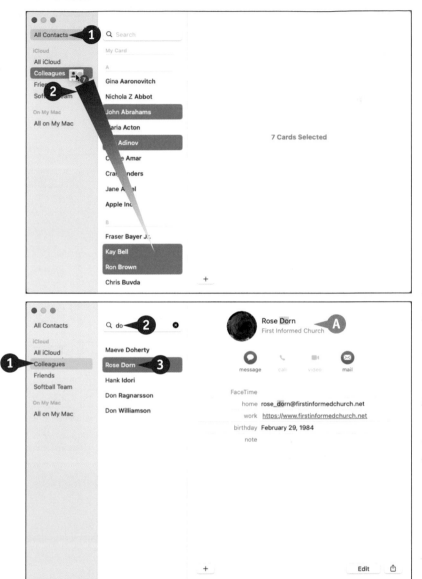

View a List or Search Within It

1 Click the list.

Contacts displays the contacts in the list.

2 To search within the list, click in the Search box and type a search term.

Contacts displays matching contacts.

3 Click the contact you want to view.

A The contact's details appear.

TIPS

How do I remove a contact from a list?
Click the list and then click the contact. Click **Edit** and select **Remove from List**. Contacts removes the contact from the list but does not delete the contact record.

How do I delete a list?
Click the list in the Lists pane so that the list is selected in blue. Click **Edit**, and then select **Delete List**. Contacts displays a confirmation message. Click **Delete**. Deleting a list does not affect the contacts it contains; the contacts remain available through the All Contacts list or any other lists to which they belong.

Create Notes

The Notes app enables you to create notes stored in an online account, such as your iCloud account, and sync them across your devices. Alternatively, you can store notes on your MacBook itself, which is useful for private notes you do not want to sync in online accounts.

The Notes app automatically saves your notes when you make changes. You can create straightforward notes in plain text, but you can also add formatting, check boxes, photos, and other items.

Open the Notes App and View an Existing Note

1 Click **Notes** (⬤) on the Dock.

Note: If Notes (⬤) does not appear on the Dock, click **Launchpad** (▦), and then click **Notes** (⬤) on the Launchpad screen.

The Notes window opens.

2 Click an existing note.

Ⓐ The contents of the note appear in the right pane.

Work with Folders

1 Click **View** on the menu bar, and then click **Show Folders**.

Ⓑ The Folders pane appears.

2 Click **New Folder** (⊕).

The New Folder dialog opens.

3 In the Name box, type the name for the folder.

Ⓒ Select **Make into Smart Folder** (☑) if you want to create a smart folder. The New Smart Folder dialog opens, enabling you to specify the details for the smart folder.

4 Click **OK**.

Notes creates the new folder, which appears in the Folders pane.

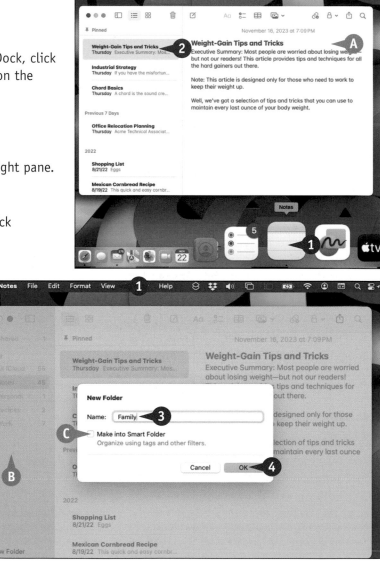

254

Create a New Note and Apply Formatting

1 In the Folders pane, click the folder in which you want to create the new note.

Note: If you do not have the Folders pane displayed, you cannot control which folder Notes creates the new note in. You can move the note to another folder later if necessary.

2 Click **Create a Note** ().

Notes creates a new blank note.

3 Type the title of the note and press Return.

Note: To set the style Notes uses for the first paragraph of a note, click **Notes** on the menu bar, and then click **Settings**. In the Settings window, click **New Notes Start With** (◆), and then click **Title**, **Heading**, or **Body**, as needed. Click **Close** (●).

4 Start typing the text of the note.

5 When you need to change the style for a paragraph, click **Styles** (Aa).

The Styles pop-up panel opens.

6 Click the style you want to apply: **Title**, **Heading**, **Subheading**, **Body**, **Monostyled**, **Bulleted List**, **Dashed List**, or **Numbered List**.

Note: Notes automatically switches to the Body style for the paragraph after a Title paragraph or a Heading or Subheading paragraph. Notes continues the Body, Monostyled, Bulleted List, Dashed List, and Numbered List styles until you change styles manually.

TIPS

Can I use styles and formatting in all my notes?
Styles are available only in notes you store in iCloud or in the On My Mac account. Basic formatting — boldface, italic, and underline — works for notes in Google, Exchange, and IMAP accounts as well as for notes in iCloud.

How do I store notes on my MacBook itself?
Click **Notes** and click **Settings**. In the Settings window, select **Enable the On My Mac account** (✓). Click **Close** (●) to close the Settings window.

continued ▶

The Notes app includes eight built-in styles that enable you to format your notes with a title, headings, body text, monostyled text, bulleted lists, dashed lists, and numbered lists. By using these styles to format notes instead of using direct formatting, such as bold and italic, you can create structured notes that you can easily use in a word processing app.

You can also create checklists — lists from which you can check off completed items — and tables with as many columns and rows as you need.

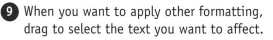

Create Notes (continued)

The Styles pop-up panel closes.

Notes applies the style to the paragraph.

7 To create a checklist, drag to select the paragraphs for the list.

8 Click **Make a checklist** (☰).

A check circle — a round check box (○) — appears before each paragraph.

Ⓐ You can click a check circle to select it (○ changes to ✓).

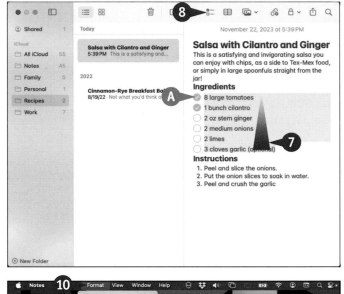

9 When you want to apply other formatting, drag to select the text you want to affect.

10 Click **Format**.

The Format menu opens.

11 Click or highlight **Font**, **Text**, or **Indentation**. This example uses **Font**.

The continuation menu opens.

12 Click the formatting you want to apply. For example, click **Italic**.

Notes applies that formatting to the text.

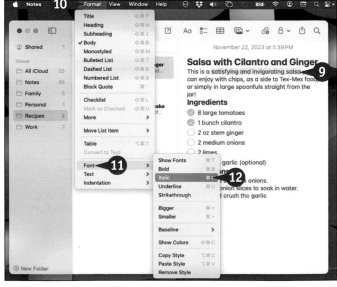

Insert a Table

1 Position the insertion point where you want the table to appear.

2 Click **Add Table** () to insert a table with two columns and two rows.

The insertion point appears in the first cell.

3 Type the content for the first cell.

4 Press **Tab** to move the focus to the next cell (not shown).

5 Type the content for the next cell.

6 Press **Control**+click **Select Column** (⬚).

The pop-up menu appears.

7 Click **Add Column After** to add a new column.

8 Continue typing the cell entries, pressing **Tab** to move to the next cell.

Note: If the current cell is the last cell in the table, pressing **Tab** automatically adds a row to the table.

B You can add a new row below the current row by pressing **Control**+clicking **Select Row** (⬚) and then clicking **Add Row Below** on the pop-up menu.

9 When you finish working with Notes, click **Close** (●) to close Notes.

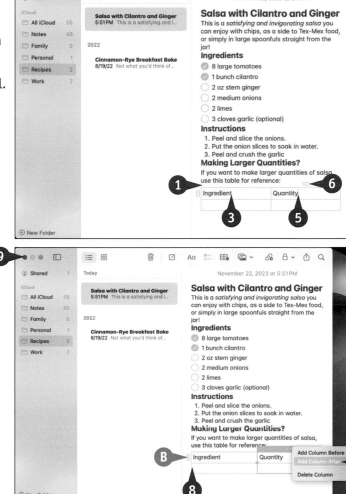

TIPS

How do I move a note from one folder or account to another?

Click **View** on the menu bar, and then click **Show Folders** to display the Folders pane. Click the current folder, and then click the note and drag it to the destination folder.

How do I delete a note?

Either click the note in the Notes list and then click **Delete** (🗑) on the toolbar, or press **Control**+click the note and then click **Delete** on the contextual menu.

Notes moves the note to the Recently Deleted folder, from which you can recover it if necessary. After 30 days, Notes permanently deletes the note.

Track Your Commitments with Reminders

The Reminders app gives you an easy way to track what you have to do and your progress on your tasks. Reminders enables you to link a reminder to a specific time, a specific location, or both. Linking a reminder to a location is useful when you sync your reminders from your MacBook with an iPhone or iPad that you carry from location to location. You can assign your reminders to different reminders lists for easy management.

Track Your Commitments with Reminders

Open Reminders and Manage Your Reminders Lists

1 Click **Reminders** ( ) on the Dock or in Launchpad to open the Reminders app.

A The sidebar on the left shows your various lists of reminders.

Note: If the sidebar does not appear, click **View** on the menu bar and click **Show Sidebar** to display it.

B The main pane shows the reminders in the selected list.

2 Click **Add List** ( ) to open the New List dialog.

3 Type the name for the new list.

4 In the Color area, click the list color.

5 In the Icon area, either click **Emoji** ( ), and then click the emoji to assign; or click **Glyph** ( ), and then click the icon to assign.

C You can click the **List Type** pop-up menu ( ) and click **Smart List** to create a smart list or **Groceries** to create a grocery list instead of a standard list.

6 Click **OK**.

Note: You can drag your lists of reminders into a different order if you want.

Note: To delete a list of reminders, press Control +click the list, click **Delete**, and then click **Delete** in the confirmation dialog.

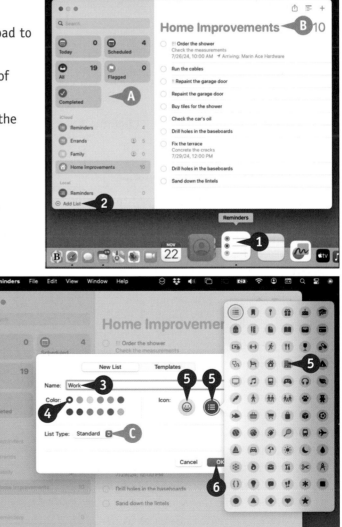

258

Create a Reminder

1 In the sidebar, click the list in which to create the new reminder.

2 Click **Add** (**+**).

A new reminder appears in the list.

3 Type the text of the reminder.

The Info button (ⓘ) appears to the right of the reminder.

4 Click **Add Date** (▦).

A pop-up panel opens.

D You can click a suggested date, such as **Today** or **Tomorrow**.

5 To choose a custom date, click **Custom**.

The calendar panel opens.

6 Double-click the date for the task.

E You can click **Add Location** (➤) to quickly add a location at which you want to receive a reminder for a task. Clicking this button produces a short list of suggested locations. To choose other locations, see the tip.

F You can click **Flag** (⚑ changes to 🚩) to flag the task. The task then appears in the Flagged list in the sidebar.

TIP

How do I link a reminder to a location?

In the pop-up panel containing the reminder's details, select **At a Location** (☑). A new section appears under At a Location. Click **Enter a Location** to display the Suggestions list; if a suggestion is suitable, click it. If not, type a contact's name or an address, and then click the correct location in the Suggestions list that appears. Click either **Getting in car** (🚗) to specify departing from the location or **Getting out of car** (also 🚗) to specify arriving at the location.

continued ▶

To find the reminders you need to work with, you can search through your reminders.

When you have completed the task for a reminder, you select the reminder's check box to mark it as complete. The reminder then disappears from your reminder lists, but you can view your completed reminders at any time by displaying the Completed list.

Track Your Commitments with Reminders (continued)

The date appears on the date button.

A You can click **Add Time** (🕐) to add a time to the task. The pop-up menu provides a list of suggested times. To choose another time, use the Info panel for the task.

B You can click **Tag** (#) to add one or more tags to the task. You can either click an existing tag in the pop-up menu or click **New Tag** and then type a new tag.

7 Click **Info** (ⓘ).

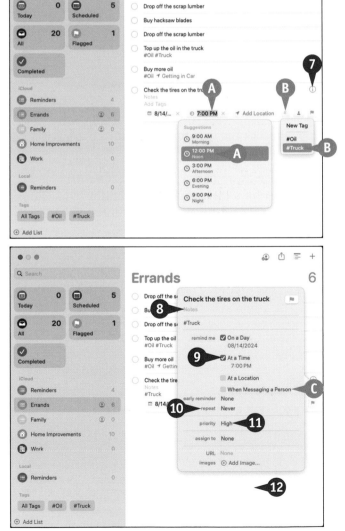

The Info panel appears.

The date and time controls appear.

8 Click **Notes** and type any notes needed for the reminder.

9 Select **At a Time** (☑) and then set the time for the reminder.

C You can select **When Messaging a Person** (☑) and then specify the appropriate contact.

10 If you want the reminder to repeat, click **repeat** and specify the frequency, such as **Every Week**.

11 Click **priority** (◇) and then click the priority: **None**, **Low**, **Medium**, or **High**.

12 Click outside the Info panel.

The Info panel closes.

Search for a Reminder

1 Click in the Search box.

Note: You do not need to select a particular list before searching, because Reminders searches across all your reminder lists.

2 Start typing your search term or terms.

D A list of matches appears.

E You can click **Clear** () to clear the search, restoring the view to the reminders list you were viewing before.

Mark a Reminder as Completed

1 Click the check circle to the left of the reminder (○ changes to ⊘).

Reminders slides the completed reminder to the top of the list and then removes it.

Note: To see the reminders you have completed, click **View** on the menu bar and click **Show Completed**.

Note: To delete a reminder, press `Control`+click it, and then click **Delete** on the contextual menu.

TIPS

How can I look at multiple reminder lists at the same time?
Double-click a list to open a new window showing that list.

How do I change the color of a reminder list?
Press `Control`+click the list to display the contextual menu and click **Show List Info** to open the Info dialog for the list. In the Color area, click the new color, and then click **OK** to close the dialog.

Make the Most of the Maps App

The Maps app enables you to pinpoint your MacBook's location by using known wireless networks. You can view your location on a road map, display a satellite picture, or view transit information for some areas. You can easily switch among map types to find the most useful one for your current needs, and you can use Maps to get directions to where you want to go.

Make the Most of the Maps App

Open Maps and Find Your Location

1 Click **Maps** (🗺️) on the Dock or Launchpad.

The Maps window opens.

2 Click **Show Your Current Location** (➤).

Ⓐ A blue dot shows your current location.

Ⓑ The scale appears in the upper-left corner of the Maps window.

Change the Map Type and Zoom In or Out

1 Click **Map Mode** (📖).

The Map Mode menu opens.

2 Click **Satellite** to switch to Satellite view.

3 Click **Zoom Out** (—) or **Zoom In** (+) to zoom in or out, as needed.

Note: You can rotate the map by placing two fingers on the trackpad and turning them. To return the map to its default northward orientation, click the orange triangle on the compass (⊙).

Ⓒ You can click **Show 3D Map** (3D) to toggle the 3D map.

Get Directions

1 Click **Directions** (⊙).

The Directions pane opens.

2 Click **Start** and type your start point.

The Suggestions panel appears.

3 Click the best suggestion.

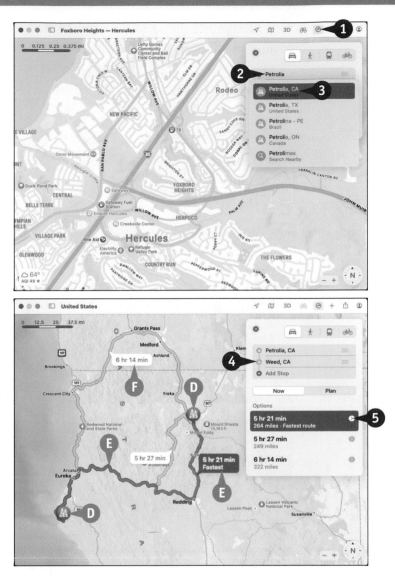

4 Click **End** and type your end point.

Maps displays suggested routes.

5 Click **Details** (⊙) for the route you want to view.

The current route's details appear in the Directions pane.

Ⓓ Maps displays symbols (🏛) to indicate your start point and end point.

Ⓔ The current route appears in darker blue.

Ⓕ You can click another route or its time box to display its details.

TIPS

How do I get directions for walking?
Click **Walk** (🚶) in the Directions pane to display the distance and time for walking the route. Be aware that walking directions may be inaccurate or optimistic. Before walking the route, verify that it does not send you across pedestrian-free bridges or through rail tunnels.

How can I send directions to my iPhone?
Click **Share** (⬆) to display the Share pop-up menu, and then click **Send to iPhone**; if you have an iPad, you can click **Send to iPad** instead. Unlock your iPhone if it is locked, and then tap the banner that appears. The Maps app on the iPhone displays the directions.

CHAPTER 10

Enjoying Music, Video, and Books

Your MacBook comes equipped with apps for enjoying music, podcasts, video, and books. The Music app enables you to import digital music, copy songs from CDs, play them back, and create playlists. The TV app enables you to watch movies, TV shows, and videos. The Podcast app lets you enjoy podcasts. The Books app enables you to build an e-book library on your MacBook and read e-books anywhere.

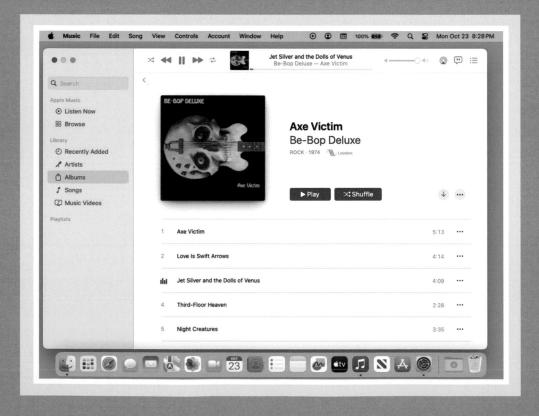

Add Your Music to the Music App

The Music app enables you to build your music library quickly by adding your existing songs to it. You can import digital audio files in formats including MP3, AAC, Apple Lossless, and WAV from your MacBook or buy songs online. If you have CDs, you can copy songs from them by using an optical drive connected to your MacBook.

When importing songs from CDs, you can choose among different settings to create files using different formats and higher or lower audio quality. The highest-quality files give the best sound but require the most space on your MacBook's disk.

Launch the Music App

Click **Music** (🎵, A) on the Dock to launch the Music app.

If the Welcome to Apple Music dialog opens, click **Start Listening** (B) to dismiss it. If the Hear About New Music First dialog then opens, offering you notifications about your favorite musicians, new features, live events, and special features, click **Enable Notifications** or **Not Now**, as appropriate.

You can then start using the Music app.

Choose the Location and Settings for Your Music Library

With the Music app active, click **Music** on the menu bar, and then click **Settings** on the Music menu to open the Settings dialog. Click **Files** (🗁, C) to display the Files tab. Look at the location (D); if necessary, click **Change** (E), click the folder you want to use, and then click **Open**.

Select **Keep Music Media folder organized** (☑, F) if you want Music to rename and organize files automatically. This behavior is usually helpful. Select **Copy files to Music Media folder when adding to library** (☑, G) to have Music copy files to the folder when adding them from other locations. This behavior is helpful for adding files from external media but creates duplicate files when adding music from your Mac's drive.

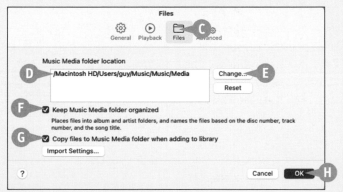

Click **OK** (H) to close the Settings dialog.

Add Digital Audio Files to the Music App

If you have digital audio files that you have not yet added to your music library, you can add them in either of these ways:

- **Music app.** Click **File** to open the File menu, and then click **Import** to open the Import dialog. Navigate to the folder that contains the files, and then click **Open**.

- **Finder.** Open a Finder window to the folder that contains the files. Drag either folders or songs from the Finder window to the Library section (I) of the sidebar in the Music app's window. When the pointer displays a green circle containing a + sign (⊕, J), drop the folders or songs.

You can also press **Control**+click a digital audio file in a Finder window, click **Open With** on the contextual menu, and then click **Music** (🎵) on the continuation menu. If the continuation menu shows "Music (default)," you can simply double-click a digital audio file. Music adds the file and starts playing it.

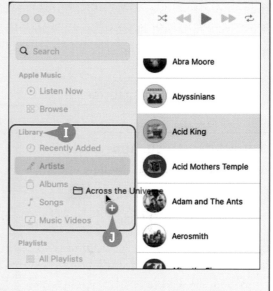

Add Songs from Your CDs to Your Music Library

To add a CD to your music library, connect an optical drive to your MacBook via USB, and then insert the CD in the optical drive. The CD (K) appears in the Devices list in the sidebar.

If a dialog opens prompting you to import the CD, click **Yes** (L). If no dialog opens, click **Import CD** (M). In the Import Settings dialog that opens, click the **Import Using** pop-up menu (🔾), and then click the encoder to use, such as **AAC Encoder**. Click the **Setting** pop-up menu (🔾), and then click the appropriate setting, such as **iTunes Plus**. Select **Use error correction when reading Audio CDs** (☑), and then click **OK** to close the dialog. After Music imports the songs, click **Eject** (⏏, N) to eject the CD.

Set Up Home Sharing

macOS, iOS, and iPadOS include a feature called *Home Sharing* that enables you to share songs and videos among your and your family's Macs, PCs, iPads, and iPhones. Home Sharing saves time and effort over copying song files manually between computers. Each family member who uses Home Sharing must have an Apple ID, such as the one created when setting up an account on Apple's iTunes Store. Anyone without an Apple ID can create one in a couple of minutes.

Set Up Home Sharing

1 Click **System Settings** (⚙) on the Dock.

The System Settings window opens.

2 Click **General** (⚙).

The General pane appears.

3 Click **Sharing** (◆).

The Sharing pane appears.

4 In the Content & Media section, set the **Media Sharing** switch to On (⬤).

The dialog for configuring sharing opens.

Ⓐ If the dialog for configuring sharing does not open automatically, click **Info** (ⓘ) to open the dialog.

5 Change the library name if necessary.

6 Select **Home Sharing** (☑).

Note: If the Enter the Apple ID Used to Create Your Home Share dialog opens, authenticate yourself by typing your email or phone number, entering your password, and then clicking **Turn On Home Sharing**.

7 Select **Devices update play counts** (☑) if you want Music to update the play counts for songs that devices play.

8 Select **Share photos with Apple TV** (☑) if you want to share your MacBook's photos with your Apple TV. Click **Choose** and use the resulting dialog to specify which photos to share.

9 Select **Share media with guests** (☑) if you want to share media files with guests on your network. You can click **Options** and choose which files to share.

10 Click **Done**.

The dialog for configuring sharing closes.

11 On the computer from which you want to browse the shared library, open the Music app and click **Source** (˅).

The Source pop-up menu opens.

12 Click the appropriate shared library.

The shared content appears.

13 Click **Show** (◉), and then click **Items not in my library**.

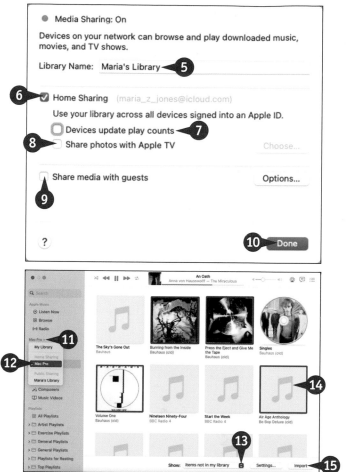

14 Select the items you want to import.

15 Click **Import**.

The Music app imports the items.

How can I make songs play at a consistent volume?

Turn on the Sound Check feature. Click **Music** on the menu bar, click **Settings**, and then click **Playback** (▶). Next, select **Sound Check** (☑), and then click **OK**.

Sound Check performs volume normalization by analyzing the songs in your library and calculating the optimum volume at which to play each song without any song being too loud or too quiet. Once you have enabled Sound Check, Music uses it for each song in your library.

Play Songs

Music makes it easy to play back your songs. You can view your music listed by songs, albums, artists, composers, or genres, which enables you to quickly locate the songs you want to hear. You can also search for songs using artist names, album or song names, or keywords.

After locating the music you want to hear, you can start a song playing by simply double-clicking it. You can then control playback by using the straightforward controls at the top of the Music window.

Play Songs

Play a Song in Songs View

1 In Music, click **Songs** (♪).

Music switches to Songs view.

The column browser appears.

Note: If the column browser does not appear, click **View**, click **Column Browser**, and then click **Show Column Browser**.

2 Click the genre.

3 Click the artist.

4 Click the album.

5 Double-click the song.

The song starts playing.

Play a Song in Albums View

1 Click **Albums** (□).

Music switches to Albums view.

2 Click the album you want to open (not shown).

The album's contents appear.

3 Double-click the song.

The song starts playing.

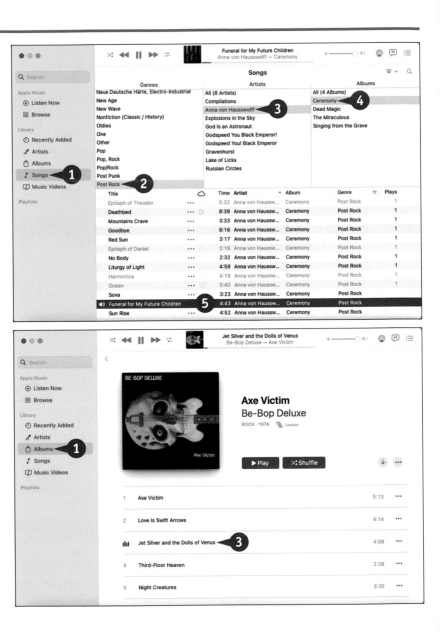

Play a Song in Artists View

1 Click **Artists** (🎤).

Music switches to Artists view.

2 Click the artist whose music you want to see.

Note: If necessary, scroll up or down to locate the album you want.

3 Double-click a song to start it playing.

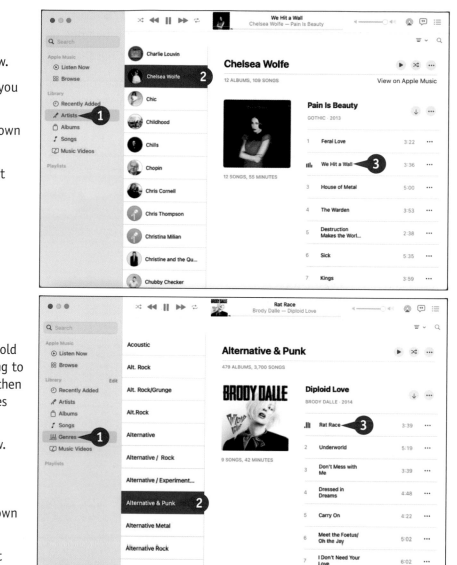

Play a Song in Genres View

1 Click **Genres**.

Note: If Genres does not appear, hold the pointer over the Library heading to make the Edit button appear, and then click **Edit**. Click **Genres** (☐ changes to ☑), and then click **Done**.

Music switches to Genres view.

2 Click the genre you want to browse.

Note: If necessary, scroll up or down to locate the album.

3 Double-click a song to start it playing.

TIP

How can I search for songs?

Click in the Search box at the top of the sidebar and start typing your search term. In the upper-right corner of the window, click **Apple Music** to search Apple Music, click **Your Library** to search your library, or click **iTunes Store** to search the iTunes Store. Music shows suggestions as you type. Click an item to go to it in your current view. On Apple Music or in your library, double-click the item to start it playing. For other actions, click the item, click **More** (•••), and then click the appropriate item on the menu that opens.

Create Playlists

The Music app enables you to create playlists that contain the songs you want in your preferred order. Playlists are a great way of getting more enjoyment out of your music. You can listen to a playlist, share it with others, or burn it to a CD for listening on a CD player.

The easiest way to create a playlist is by selecting some suitable songs and then giving the New Playlist from Selection command. You can then change the songs' order and add other songs as needed.

Create Playlists

1 In the Music app, select one or more songs you want to put into a new playlist.

Note: Click the first song you want to select, and then press ⌘+click each other song.

2 Click **File**.

The File menu opens.

3 Click **New**.

The New submenu opens.

4 Click **Playlist from Selection**.

The Playlists screen appears, showing the playlist with a default name, such as Playlist, which is selected so you can type over it.

5 Type the name for the playlist and then press **Return**.

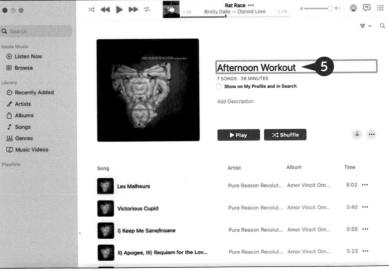

The playlist appears.

6 Select **Show on My Profile and in Search** (☑) if you want to make the playlist public, causing it to appear on your Apple Music profile and in search results.

7 Drag the songs in the playlist into the order you want.

8 Click **Add Description**.

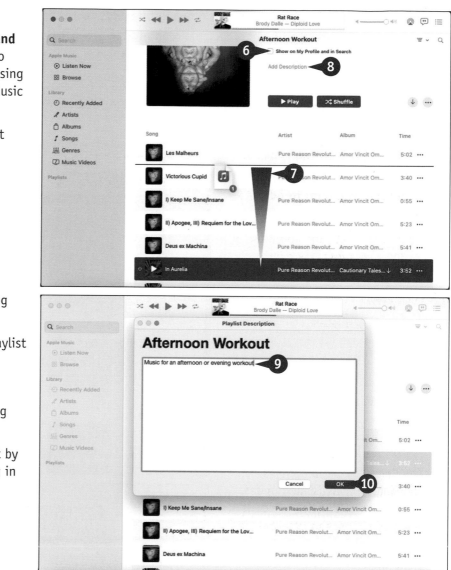

The Playlist Description dialog opens.

9 Type a description for the playlist to help you identify it.

10 Click **OK**.

The Playlist Description dialog closes.

You can now play the playlist by double-clicking the first song in the playlist.

TIP

How can I keep my playlists organized?

You can organize your playlists into playlist folders. Click **File**, click **New**, and then click **Playlist Folder**. Music creates a folder and displays an edit box around the name. Type the name for the folder and press Return to apply it. You can then drag playlists to the folder. Click **Expand** (>) to expand a folder and reveal its playlists; click **Collapse** (∨) to collapse the folder and hide its playlists.

Create Smart Playlists

Instead of creating playlists manually by adding songs to them, you can have the Music app's Smart Playlists feature create playlists automatically for you. A *Smart Playlist* is a playlist Music builds based on criteria you specify. You can either freeze the Smart Playlist once you have created it or set Music to update the Smart Playlist automatically as your music library changes.

To create a Smart Playlist, you set up the criteria, also called *rules*, and name the playlist. Music then adds content to the playlist for you.

Create Smart Playlists

1 In the Music app, click **File**.

The File menu opens.

2 Click **New**.

The New continuation menu opens.

3 Click **Smart Playlist**.

Note: You can also start a Smart Playlist by pressing 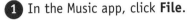.

The Smart Playlist dialog opens.

4 Click the first pop-up menu (⬍) and select the item for the first condition — for example, **Genre**.

5 Click the second pop-up menu (⬍) and select the comparison for the first condition — for example, **contains**.

6 Click the text field and type the text for the comparison — for example, **Rock** — making the condition "Genre contains Rock."

7 To add another condition, click **Add** (⊕), displaying a second line of controls.

8 Click **all/any** (⬍) and select **any** to match any of the rules or **all** to match all the rules.

9 Set up the second condition by repeating steps **4** to **6**.

Note: You can add as many conditions as you need to define the playlist.

Ⓐ You can limit the playlist by selecting **Limit to** (☑) and specifying the limit.

10 Select **Live updating** (☑) if you want Music to update the playlist automatically.

11 Click **OK**.

Music adds the Smart Playlist to the Playlists section of the Source list.

Ⓑ An edit box appears around the suggested name.

12 Type the name for the Smart Playlist, and then press Return.

Music applies the name to the playlist.

TIPS

How do I produce a Smart Playlist that is the right length for a CD?
In the Smart Playlist dialog, select **Limit to** (☑). Click the left pop-up menu (⬍) and select **minutes**, and then set the number before it to **74** or **80**, depending on the capacity of the CD.

What does Match Only Checked Items do?
Select **Match only checked items** (☑) to restrict your Smart Playlist to songs whose check boxes are selected. This means you can uncheck the check box for a song to prevent it from appearing in your Smart Playlists. If the check boxes do not appear, click **Music**, click **Settings**, and click **General** (⚙) to display the General settings pane. Select **List view checkboxes** (☑), and then click **OK**.

Listen to Apple Radio and Internet Radio

The Music app enables you to listen to online radio stations. The Apple Radio feature comes set to access a selection of stations on demand, which means you can pause the radio stream and resume it from the same place. You can skip some songs if you do not want to listen to them, and you can create custom stations. To listen to Apple Radio, you must sign in to the Apple Music service.

You can also use Music to access radio stations that broadcast across the Internet in real time. When listening to such stations, you cannot pause the content or skip songs.

Listen to Apple Radio and Internet Radio

Listen to a Radio Station

1. In Music, click **Radio** (((•))).

   The Radio screen appears.

2. Position the pointer over the station you want to listen to.

   The Play button (▶) appears.

3. Click **Play** (▶).

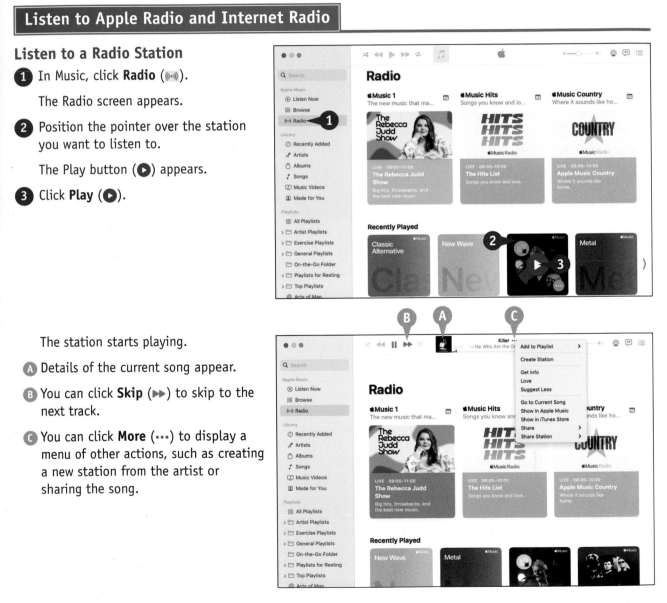

The station starts playing.

Ⓐ Details of the current song appear.

Ⓑ You can click **Skip** (▶▶) to skip to the next track.

Ⓒ You can click **More** (•••) to display a menu of other actions, such as creating a new station from the artist or sharing the song.

④ To explore the selection of genres available, scroll down to the bottom of the screen.

The More to Explore section appears.

⑤ Click the genre you want to explore.

The genre appears.

⑥ Position the pointer over the station you want to listen to.

The Play button (▶) appears.

⑦ Click **Play** (▶).

The station starts playing.

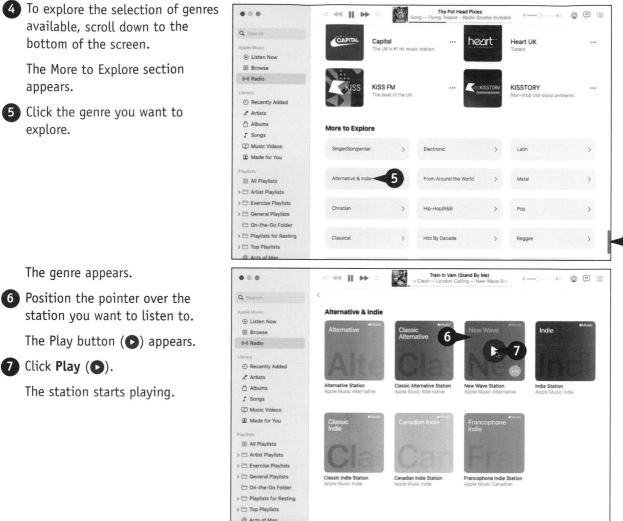

How can I listen to an Internet radio station that does not appear on Music's list?
Find out the URL of the station's audio stream by consulting the station's website. In Music, click **File** and select **Open Stream URL** to display the Open Stream dialog. Type or paste the URL into the dialog and then click **OK**. Music starts playing the radio station's audio stream.

Enjoy Podcasts

A *podcast* is an audio or video file that you can download from the Internet and play on your MacBook or a digital player like the iPhone, iPad, or iPod. The Podcasts app enables you to access the Apple Podcasts service. Apple Podcasts makes a wide variety of podcasts available. You can either download a single podcast episode or subscribe to a podcast so that Podcasts automatically downloads new episodes for you.

Enjoy Podcasts

1 Click **Launchpad** (▦) on the Dock.

The Launchpad screen appears.

2 Click **Podcasts** (🎙️).

The Podcasts app opens.

3 Click **Charts** (☰).

The Top Charts section appears.

4 Click **All Categories**.

The Categories list appears.

5 Click the category of podcasts you want to view, such as **Business**.

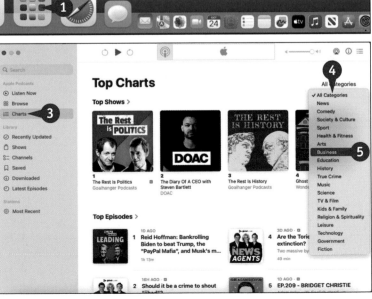

Podcasts shows the top charts for the category you clicked.

6 Click the podcast you want to view.

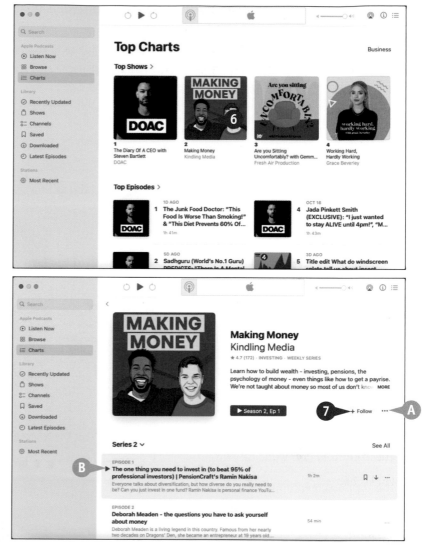

A Click **More** (•••) to display other actions, such as reporting a concern and sharing the podcast.

B To start an episode playing immediately, move the pointer over the episode, and then click **Play** (▶).

7 Click **Follow** if you want to subscribe to the podcast.

Podcasts subscribes you to the podcast and downloads the most recent episode.

TIP

How do I listen to or watch the podcasts I have subscribed to?
The Podcasts app stores the podcasts you have subscribed to in your library. The Library section of the sidebar enables you to view the library's contents in different ways. For example, click **Downloaded** (⊕) to see the episodes you have downloaded; click **Latest Episodes** (⊙) to see the podcasts that have new episodes; or click **Shows** (▢) to browse the shows you have subscribed to. Click an episode to display it, and then click **Play Episode**.

Play Videos with the TV App

The TV app enables you to play videos, TV shows, and movies. You can buy and watch TV shows and movies from Apple's services or export files of your own movies from iMovie or other applications. After adding videos to the TV app, you access them by using the tabs at the top of the application window. You can download files and watch them offline.

You can watch video content full screen or as a picture-in-picture floating window. You can also play videos from your MacBook to a TV connected to an Apple TV.

Play Videos with the TV App

1 Click **TV** (📺) on the Dock.

Note: If TV (📺) does not appear on the Dock, click **Launchpad** (⊞) to display the Launchpad screen, and then click **TV** (📺).

The TV app opens.

2 In the Library section of the sidebar, click the library category you want to view: **Recently Added** (🕐), **Movies** (🎬), **TV Shows** (🖥), **Home Videos** (🎥), **Family Sharing** (👥), or **Genres** (🎬).

This example uses **Home Videos** (🎥).

The category you clicked appears.

3 Click the item you want to view.

The item's screen appears.

4 Click the thumbnail for the item you want to play.

280

The TV app starts playing the movie full screen.

5 Move the pointer over the video.

The pop-up controls appear.

6 Use the pop-up controls in the middle of the screen to control playback.

A You can click **Picture-in-Picture** (⊡) to switch the video to a small floating window.

Note: From the picture-in-picture window, click **Expand** (⊡) to expand the video to a larger window. You can then click **View** and click **Enter Full Screen** or press ⌘+ Control + F to switch to full screen.

7 Press Esc to stop viewing the video (not shown).

 TIP

How do I play a video from the Music app on the TV connected to my Apple TV?
First, make sure your MacBook is connected to the same network as the Apple TV via either a wireless network or a wired network. You can then click **AirPlay** (◎) to the right of the volume control in the TV app to display the AirPlay menu. Click the Apple TV's name on the menu. The TV app sends the video output to the Apple TV, which displays it on the TV's screen.

Read Books

macOS includes the Books app, which enables you to enjoy electronic books, or *e-books*, on your MacBook. Using Books, you can read e-books that you already have on your MacBook, download free or paid-for e-books from online stores, or read PDF files stored on your MacBook.

Before you can read e-books that you have on your MacBook, you must add them to Books. You can then open a book and start reading it.

Read Books

Add Your E-Books to Books

Note: This method shows you how to add e-books not purchased from Apple. Any books you have bought via Books appear automatically in the Books app when you sign in using your Apple ID.

1 Click **Books** (📖) on the Dock.

Note: If Books (📖) does not appear on the Dock, click **Launchpad** (𝌆) on the Dock and then click **Books** (📖) on the Launchpad screen.

The Books window opens.

2 Click **File**.

The File menu opens.

3 Click **Import**.

The Import dialog opens.

4 Navigate to the folder that contains the books.

5 Select the books you want to add to Books.

6 Click **Import**.

The Import dialog closes.

The books appear in Books.

Read E-Books

1 Click **Books** (📖) on the Dock.

The Books window opens.

2 In the sidebar, click the button for the view by which you want to browse.
For example, click **Books** in the Library section to view the books in your library.

A You can search for a book by clicking in the Search box and typing keywords.

3 Double-click the book you want to open.

The book opens in a new window.

4 Press ➡ to display the next page or ➡ to display the previous page (not shown).

Note: You can also swipe left or right on the trackpad to change pages.

B You can click **Appearance** (AA) to adjust the appearance of the page and the text.

5 When you finish reading, click **Close** (⬤).

The book closes, and your library appears.

TIP

How can I add my Kindle e-books to Books?
As of this writing, Books cannot display books in Amazon's proprietary Kindle format. To read Kindle books on your MacBook, download the free Kindle app from the App Store and log in to the Kindle service using the e-mail address and password you have registered with Amazon.

CHAPTER 11

Making the Most of Your Photos

macOS includes Photos, a powerful but easy-to-use application for managing, editing, and enjoying your photos. You can import photos from your digital camera, phone, or tablet; crop them, straighten them, and improve their colors; and use them in albums, slide shows, or e-mail messages.

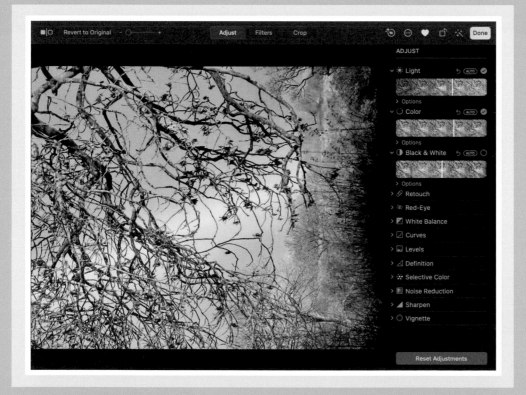

Import Photos

Photos enables you to import photos directly from a wide range of digital cameras, phones, and tablets, including the iPhone, iPad, and iPod touch. Photos normally recognizes a camera automatically when you connect it to your MacBook and switch it on.

Alternatively, you can remove the digital camera's memory card and insert it in a memory card reader connected to the MacBook; your camera's manufacturer may recommend this method. Another possibility for some cameras and phones is to import photos via Bluetooth File Exchange.

Import Photos

1 Connect your digital camera or device to your MacBook via USB (not shown).

2 Turn on the digital camera or device (not shown).

Note: Some digital cameras turn on automatically when you connect them to a powered USB port, but most cameras need to be turned on manually. Phones and tablets usually remain on and wake from sleep.

Photos opens and displays the contents of the digital camera or device.

Ⓐ If Photos does not open automatically, click **Photos** (🌸) on the Dock, click the device — such as **Chrome Ax** — in the Devices list, and then click **Open Photos** (☐ changes to ☑).

Ⓑ You can drag the **Zoom** slider to zoom in or out on the thumbnails.

Note: Scroll up and down as needed by swiping or dragging two fingers on the trackpad.

3 Click each photo you want to import, placing a check mark (☑) on it. You can drag across multiple photos to select them.

Note: To select a range of photos, click the first photo, and then press Shift +click the last photo.

Note: If you want to import all the new photos, you need not select any.

④ Click **Album** (▾) and then click the destination. Your choices are **None** to import the photos to your library, **New Album**, or an album in the My Albums list.

⑤ Click **Import Selected** to import the photos you selected. Click **Import All New Items** to import all the new photos.

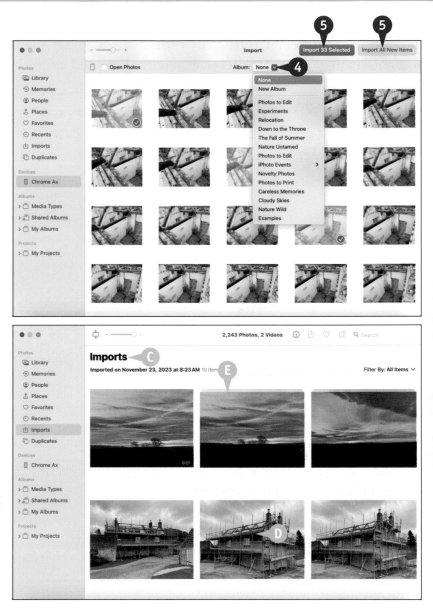

Photos copies the photos from the digital camera or device to your MacBook.

Ⓒ Photos displays the Imports screen, showing the photos you have just imported.

Ⓓ An individual photo appears as a thumbnail.

Ⓔ Photos shot in a burst appear as a stack.

Note: For a digital camera, click **Eject** (⏏) next to the camera's name in the Devices list to eject the camera before you disconnect it.

⑥ Disconnect the camera or device from your MacBook (not shown).

TIP

What is the Filter By pop-up menu for?
The Filter By pop-up menu enables you to switch between displaying all items — click **All Items** on the menu — and displaying only specific types of items. For example, you can click **Photos** (▢) to display only photos, click **Videos** (▣) to display only videos, click **Edited** (≞) to display only items you have edited, or click **Favorites** (♡) to display only items you have marked as favorites.

Browse Your Photos

To locate the photos you want to view and work with, you browse your photos. Photos enables you to browse your photos easily: You first select a category in the sidebar on the left side of the Photos window and then use the main part of the window to view the photos in the source.

If you have just imported photos, the best way to begin browsing is by viewing the Imports category, which contains the photos you have imported.

Browse Your Photos

Open the Photos App and Browse by Albums

1 Click **Photos** (🌸) on the Dock.

Photos opens.

Ⓐ You can click **Aspect Ratio** (⇡ or ⇣) to switch the thumbnails between square (⇡) and their full aspect ratios (⇣).

Ⓑ You can click **Imports** in the Library section of the sidebar to see photos you have imported.

Note: If the Albums list is hidden, move the pointer over the Albums heading, making the Expand button appear, and then click **Expand** (>).

2 Click **Expand** (> changes to ⌄) next to My Albums.

The Albums list appears.

3 Click the album you want to view.

Note: Scroll up and down as needed by swiping or dragging two fingers on the trackpad.

Ⓒ You can move the pointer over a photo and click **Add to Favorites** (🖤 changes to 🤍) to add the photo to your Favorites.

4 Double-click the photo you want to view.

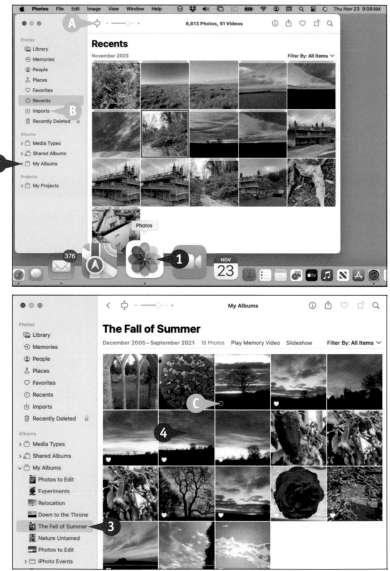

288

Photos displays the photo.

Ⓓ You can press Control +click the photo to display the contextual menu.

Ⓔ You can click **Share** to display a panel of options for sharing the photo.

Ⓕ You can click **Delete 1 Photo** to delete this photo.

⑤ Swipe left with two fingers on the trackpad to display the next photo, or swipe right with two fingers to display the previous photo.

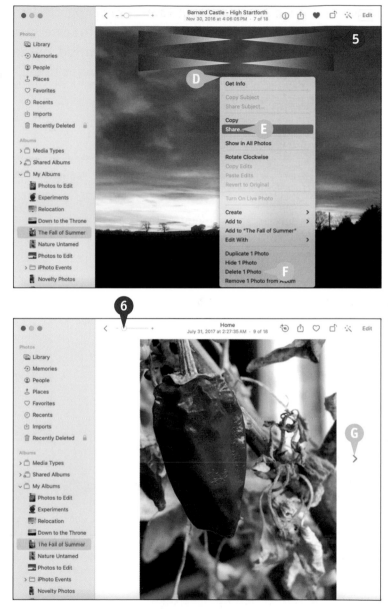

The next photo or the previous photo appears, depending on which way you swiped.

Note: You can also press ← to display the previous photo or → to display the next photo.

Ⓖ You can also switch photos by moving the pointer to the left side of the photo and clicking **Previous** (<) or moving the pointer to the right side and clicking **Next** (>).

⑥ Drag the **Zoom** slider to zoom in.

TIP

Is there another way to navigate the Photos app?
When viewing a photo, you can click **View** on the menu bar and click **Show Thumbnails** to display a bar of thumbnails across the bottom of the window. You can then click a thumbnail to display that photo. This can be a quick way to navigate large collections of photos.

continued ▶

The Photos app automatically organizes your photos into groups called Years, Months, and Days. To access these groups, you select the Photos category in the Library section of the sidebar. You can then click the Years button, the Months button, or the Days button on the bar across the top of the Photos window to display the groups you want to see, and then double-click a group to open it.

Browse Your Photos (continued)

Photos zooms in on the middle of the photo.

Note: You can also zoom in by placing two fingers, or your finger and thumb, together on the trackpad and then moving them apart. To zoom out, place two fingers, or your finger and thumb, apart and then pinch inward.

7 Drag with two fingers as needed to pan around the photo after zooming in.

8 Click **Back** (<).

The album appears again.

Browse Photos by Years, Months, and Days

1 In the Photos section of the sidebar, click **Library** (⬚).

The Photos pane appears.

2 Click **Years**.

The Years pane appears.

3 Double-click the photo thumbnail for the year you want to view.

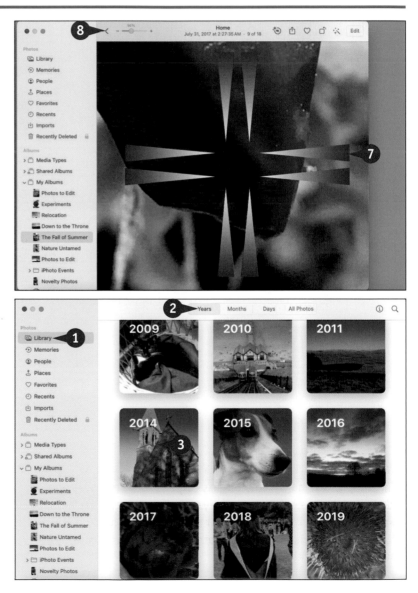

The Months pane for the year appears, showing larger thumbnails that represent the months.

④ Double-click the photo thumbnail for the month you want to view.

The Days pane for the month appears.

⑤ Double-click the photo thumbnail for the day you want to view.

What is the People item in the Photos category?
The People album contains photos of people whom you have identified to Photos. Photos analyzes the faces and attempts to use facial recognition on other photos. You can double-click a person to view all the photos in which the Photos app has identified them.

What is the Memories item in the Photos category?
The Memories album contains Memories, groups of photos that Photos automatically selects based on locations and dates.

continued ▶

The Photos app enables you to browse your photos by the places in which you took them or the places with which you tagged them. Browsing by places is helpful when you need to find photos taken at a particular place without having to search by dates. Photos shows only the main places when the map is zoomed out, but displays more places the further you zoom in.

Browse Your Photos (continued)

The photo opens.

Ⓐ You can click **Back** (<) to return to the Days screen.

Note: You can swipe left or right with two fingers to display the next photo or the previous photo.

Browse Your Photos by Places

Note: If the Photos list is collapsed, move the pointer over the Photos heading, making the disclosure triangle (>) appear, and then click **Expand** (> changes to ∨).

① In the Photos section of the sidebar, click **Places** (⚓).

The Photos app displays a map showing the places in which your photos were taken.

② Click **Zoom In** (+) one or more times to zoom in.

Ⓑ You can click **Zoom Out** (−) to zoom out.

Note: You can also zoom in by pinching apart on the trackpad or zoom out by pinching together.

Ⓒ You can click **Satellite** to switch to the photographic Satellite View.

Ⓓ You can click **Grid** to switch to Grid View.

③ Click the thumbnail for the photo or group of photos you want to see.

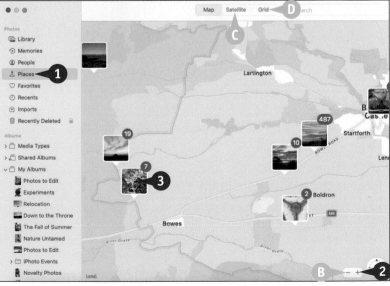

The photos taken in that place appear.

4 Double-click the photo you want to view.

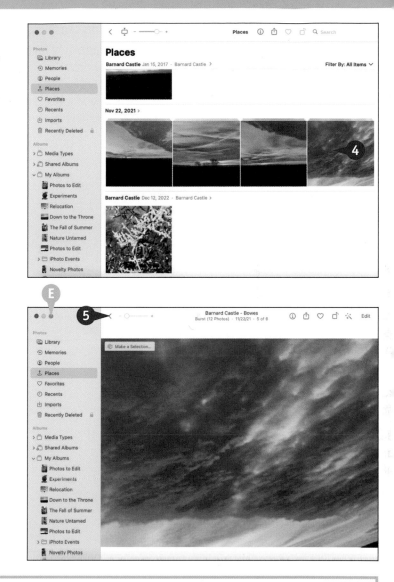

The photo opens.

E You can click **Zoom** (⬤) to expand the Photos window to full screen so that you can examine the photo more closely.

5 Click **Back** (‹).

The photos in the place appear again.

6 Click **Back** (‹) again (not shown).

The map appears again.

TIP

How do I add location information to a photo?

Open the photo and click **Info** (ⓘ) to display the Info pane. Click **Assign a Location** and start typing the address or other identifying information. In the pop-up menu of results, click the correct information. A map appears with a pin showing the location. If necessary, press `Shift` + drag the pin to adjust the location. Click **Close** (⬤) to close the Info pane.

To hide a photo's location, open the photo, click **Image** on the menu bar, click or highlight **Location**, and then click **Hide Location**. From the Location continuation menu, you can also click **Revert to Original Location** to revert the photo to its original location, or click **Copy Location** to copy the location so you can paste it into another photo.

Select Photos from Bursts

Many digital cameras and devices, including most models of iPhone, can take bursts of photos. Bursts are great when you are trying to capture facial expressions, live action, or other unrepeatable moments.

The Photos app enables you to browse the photos you have taken in bursts. From a burst, you can choose which photos to keep as individual photos; you can also choose whether to keep the rest of the burst photos or delete them.

Select Photos from Bursts

1 In Photos, navigate to the burst. To see all bursts, expand **Albums**, expand **Media Types**, and then click **Bursts** (📷).

Note: If you have added the burst recently, click **Recents** (🕐).

Note: Scroll up and down as needed by swiping or dragging two fingers on the trackpad.

Ⓐ Each burst appears as a stack of photos.

2 Double-click the burst you want to open.

Photos displays the photo at the top of the stack in the burst.

Ⓑ The Burst indicator shows you how many photos the burst contains.

3 Click **Make a Selection**.

Note: The Make a Selection prompt appears only on a burst of photos, not on individual photos.

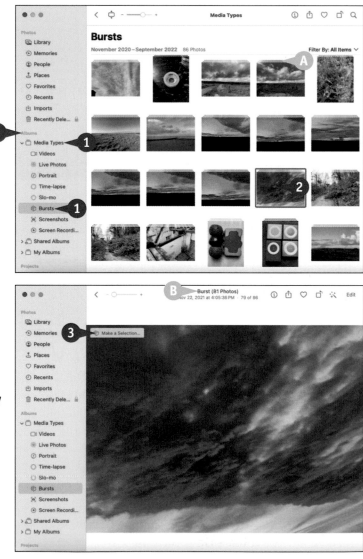

The Make a Selection pane appears.

C You can swipe left or right with two fingers, or press ➡ or ⬅ to display the next photo or previous photo.

D The inverted triangle indicates the thumbnail for the current photo.

4 Click the thumbnail for the photo you want to view.

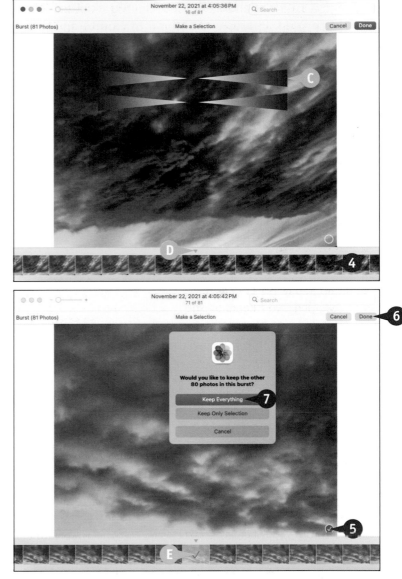

The photo appears.

5 Select the selection circle (✓) if you want to keep the photo.

E A check mark appears on the thumbnail for each photo you select.

6 When you finish selecting photos from the burst, click **Done**.

The Would You Like to Keep the Other Photos in This Burst? dialog opens.

7 Click **Keep Everything** or **Keep Only Selection**, as needed. Click **Cancel** to return to the Make a Selection pane.

TIP

How do I take a burst of photos on my iPhone?
In the Camera app, make sure that Photo Mode is selected so that PHOTO appears in yellow. Drag the Shutter icon toward the last item's thumbnail to start taking a burst — in other words, with the iPhone held in portrait orientation, drag the Shutter icon to the left. If you find this maneuver awkward, tap **Settings** (⚙), tap **Camera** (📷), and then set the **Use Volume Up for Burst** switch to On (⚪). You can then hold down **Volume Up** to shoot a burst.

Crop a Photo

To improve a photo's composition and emphasize its subject, you can crop off the parts you do not want to keep. Photos enables you to crop to any rectangular area within a photo, so you can choose exactly the part of the photo that you need. You can either constrain the crop area to a specific aspect ratio or crop freely. Constraining an area to specific dimensions is useful for producing an image with a specific aspect ratio, such as 3 × 5 or the ratio of your MacBook's display.

Crop a Photo

1 In Photos, open the photo you want to edit.

2 Click **Edit**.

Photos opens the photo for editing and displays the editing tools.

3 Click **Crop**.

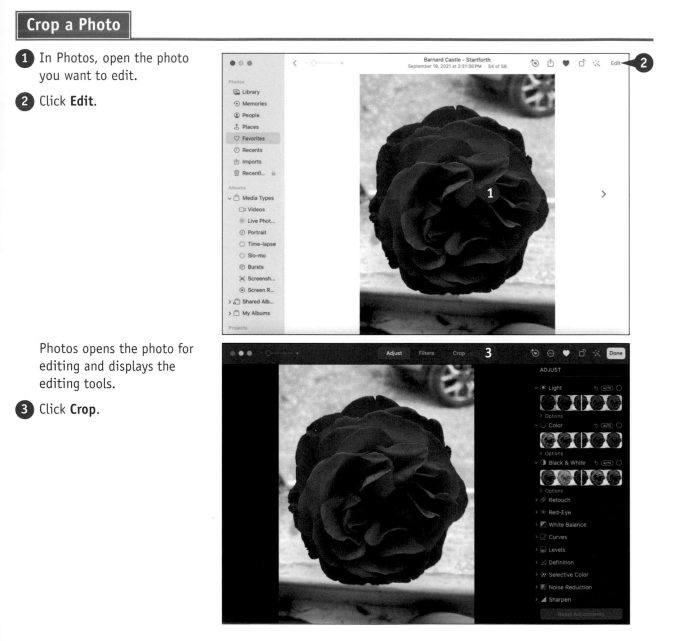

Photos displays the cropping tools.

4 To crop to specific proportions or dimensions, go to the Aspect (▥) list and click the aspect ratio you want to use, such as **Square**.

Ⓐ You can click **Flip** (▨) to flip the photo horizontally. Press **Option** (▨ changes to ▶) and then click **Flip** (▶) to flip the photo vertically.

The crop box changes to show the aspect ratio you chose.

5 Drag the corner handles to crop to the area you want.

6 If necessary, click inside the cropping rectangle and drag the picture to change the part shown.

Ⓑ You can click **Revert to Original** at any point to undo all the changes you have made to the photo.

7 Click **Done**.

Photos crops the picture to the area you chose.

Photos hides the editing tools again.

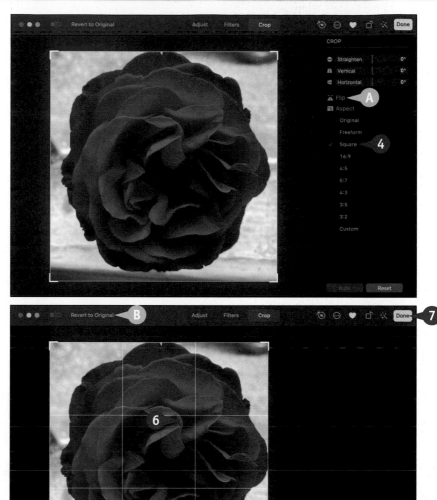

TIPS

I cropped off the wrong part of the photo. How can I get back the missing part?
Open the photo, click **Edit**, and then click **Revert to Original**.

How can I crop a photo evenly around its center?
Hold **Option** as you drag the cropping handles. Holding **Option** makes Photos adjust the crop box around the center of the photo.

Rotate or Straighten a Photo

With digital cameras, and especially with phones and tablets, you can easily take photos with the device sideways or the wrong way up. The Photos app enables you to rotate a photo easily by 90, 180, or 270 degrees to the correct orientation.

Photos also enables you to straighten a photo by rotating it a few degrees clockwise or counterclockwise. To keep the straightened picture in its current aspect ratio, Photos automatically crops off the parts that no longer fit.

Rotate or Straighten a Photo

Rotate a Photo

1 In Photos, press `Control` + click the photo you want to rotate.

The contextual menu opens.

2 Click **Rotate Clockwise**.

Note: Press `Option` and then click **Rotate Counterclockwise** to rotate the photo counterclockwise.

Photos rotates the photo 90 degrees clockwise.

Note: If you need to rotate the photo further, repeat the move.

Straighten a Photo

1 In Photos, open the photo you want to straighten.

2 Click **Edit**.

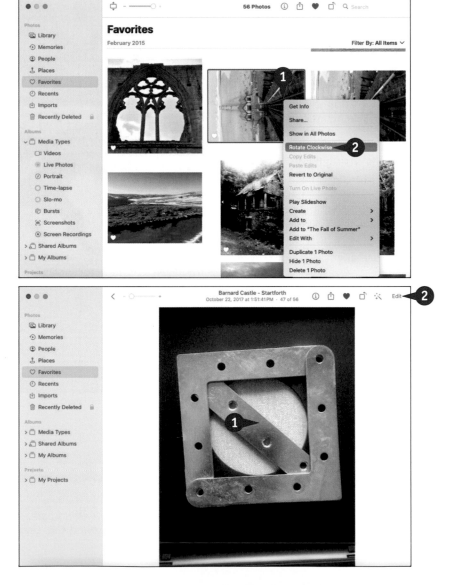

Photos opens the photo for editing and displays the editing tools.

3 Click **Crop**.

Ⓐ You can click **Rotate** (🔄) to rotate the photo 90 degrees counterclockwise. To rotate the photo 90 degrees clockwise, press `Option` (🔄 changes to 🔄) and then click **Rotate** (🔄).

Photos displays the cropping and straightening tools.

4 Drag the **Straighten** slider to straighten the photo.

Note: Use the lines in the straightening grid to judge when lines in the picture have reached the desired horizontal or vertical position.

5 Click **Done**.

Photos applies the straightening.

Photos hides the editing tools again.

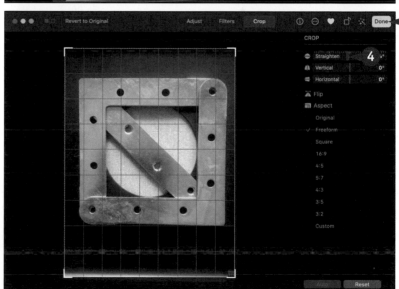

TIP

What does the Auto button in the Crop pane do?

Click the Auto button to apply automatic straightening to the photo you are editing. This feature analyzes the horizontal and vertical lines in the photo and attempts to apply straightening based on what it finds. For photos that have obvious orientation problems, this feature can work well. For other photos, such as those where you need to use the background rather than a foreground object to determine what should be horizontal and what should be vertical, you may do better to apply straightening manually.

Improve a Photo's Colors

Photos includes powerful tools for improving the colors in your photos. If a photo is too light, too dark, or the colors look wrong, you can use these tools to make it look better.

Usually, the best way to start is to use the Auto Enhance tool, which boosts flat colors while muting overly bright ones. If Auto Enhance does not give you the results you need, you can use the Adjust tools to edit settings such as exposure, contrast, and saturation.

Improve a Photo's Colors

Quickly Enhance the Colors in a Photo

1 In Photos, open the photo you want to enhance.

2 Click **Auto Enhance** (🪄 changes to 🪄).

Photos adjusts the exposure and enhances the colors.

Note: If the Auto Enhance tool does not improve the photo, click **Auto Enhance** again (🪄 changes to 🪄) to remove the enhancement, and use the Adjust tools instead, discussed next.

Using the Adjust Tools

1 Open the photo you want to adjust.

2 Click **Edit**.

Photos opens the photo for editing and displays the editing tools.

③ Click **Adjust**.

The Adjust tools appear.

Note: If the set of tools you want to appear is collapsed, move the pointer over its heading and click **Expand** (▶) to expand it.

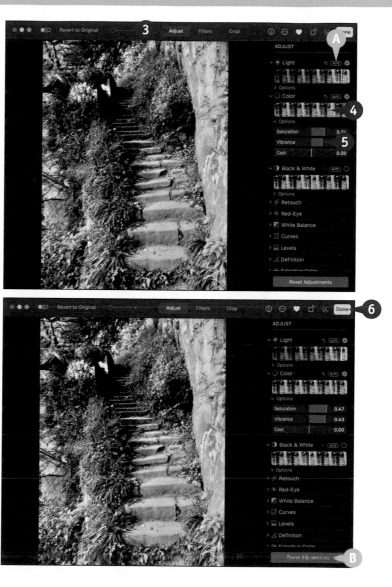

Ⓐ You can click **Auto** to have Photos adjust the group of settings automatically.

④ Drag the group's slider to adjust all the settings using preset balances.

⑤ Drag an individual slider to adjust that setting alone.

Ⓑ You can click **Reset Adjustments** to undo your adjustments.

Note: If you need to make the same adjustments to other photos, press Control+click the photo and click **Copy Adjustments**. You can then open each photo for editing, press Control+click the photo, and click **Paste Adjustments** to apply the copied adjustments.

⑥ When you finish working on the colors, click **Done**.

Photos applies the changes to the photo.

Photos hides the editing tools again.

TIP

What other adjustments can I make using Photos?
You can make a wide range of adjustments by using the tools in the lower section of the Adjust pane. The tools include Retouch for removing blemishes; White Balance for adjusting misrepresented white tones; Levels for changing red-green-blue levels, white point, and black point; Curves for adjusting tonal range; Selective Color for adjusting only the color of your choice; Noise Reduction for reducing *noise* or graininess in low-light photos; Sharpen for increasing definition; and Vignette for fading out the corners of the photo to emphasize the subject.

Add Filters to Photos

Photos includes ten preset filters that you can quickly apply to change a photo's look and add life and interest to it. For example, you can change a color photo to black-and-white, boost or fade the color, or apply an instant-camera filter.

To add filters, you open the photo for editing and display the Filters panel. You can then experiment with the available filters to see which filter works best. After applying a filter, you can adjust its intensity if necessary.

Add Filters to Photos

1 In Photos, open the photo to which you want to apply a filter.

2 Click **Edit**.

Photos opens the photo for editing and displays the editing tools.

3 Click **Filters**.

The Filters pane appears.

④ Click the filter you want to apply. This example uses the Vivid Warm filter.

Ⓐ Photos applies the filter at full intensity — 100 out of 100 — at first.

Note: The Photo Booth app provides a similar but more extensive collection of filters.

The photo takes on the filter.

Ⓑ If you want to reduce the filter's intensity, drag the slider to the left.

⑤ When you are satisfied with the filter, click **Done**.

Photos applies the filter to the photo.

Photos hides the editing tools again.

TIP

How can I remove the filter from a photo?

To remove the filter you have applied to a photo, open the photo for editing, click **Filters**, and then click **Original** at the top of the Filters pane.

Create Photo Albums

When you want to assemble a custom collection of photos, you create a new album. You can then add to it exactly the photos you want from any of the sources available in Photos. After assembling the collection of photos, you can arrange them in your preferred order.

Photos can also create *Smart Albums* that automatically include all photos that meet the criteria you choose. Photos updates Smart Albums automatically when you download photos that match the criteria.

Create Photo Albums

① In Photos, move the pointer over the My Albums item in the Albums section of the sidebar.

The Add (⊕) button appears.

② Click **Add** (⊕).

The Add pop-up menu opens.

③ Click **Album**.

Note: To add photos to an existing album, select the photos, click **Image** on the menu bar, highlight **Add to** without clicking, and then click the album on the Add To continuation menu.

Ⓐ Photos creates a new album, assigning it a provisional name — in this example, Recents, because the Recents category was selected — and placing an edit box around that name in the sidebar.

④ Type the name for the new album and press Return.

Note: If Photos adds any photos to the album, delete them.

⑤ In the sidebar, click the photo collection from which you will add photos to the new album. This example uses the **Library** (🖼) category.

The photo collection you clicked appears.

6 Click the first photo you want to add.

A blue outline appears around the photo.

7 Press ⌘+click each other photo you want to add.

A blue outline appears around each photo you click.

8 Drag the photos to the new album in the sidebar.

9 Click the new album.

The photos in the album appear.

10 To change the order of the photos, click a photo and drag it to where you want it.

Photos rearranges the photos.

11 Click **Back** (‹) when you want to return to the Albums pane.

What is a Smart Album, and how do I create one?

A Smart Album is an album based on criteria you choose. For example, you can create a Smart Album of photos with the keyword "family" and a rating of four stars or better. Photos then automatically adds each photo that matches those criteria to the Smart Album. To create a new Smart Album, click **File** and select **New Smart Album**, and then set your criteria in the New Smart Album dialog. Click **Add** (+) to add another row of criteria to the Smart Album.

Create and Play Slide Shows

One of the best ways to enjoy your photos and share them with others is to play a slide show. Photos enables you to play an instant slide show by selecting the collection of photos you want to view, clicking **File** to open the File menu, and then clicking **Play Slideshow**. When you need greater control over the slide show, you can create a saved slide show with custom music and effects. You select photos, create the slide show, arrange the photos into your preferred order, and save the show so you can run it when needed.

Create a Saved Slide Show

Start by choosing the photos to include in the slide show. Click the photo collection from which you first want to select photos, click the first photo you want to use, and then ⌘+click each of the other photos.

Once you have selected the photos, click **File** (A) to open the File menu, highlight **Create** (B) without clicking, click **Slideshow** (C), and then click **Photos** (D).

In the Add Photos to Slideshow dialog that opens, type the slide show name in the Slideshow Name box (E), and then click **OK** (F).

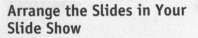

Arrange the Slides in Your Slide Show

Photos creates the slide show, adds it (G) to the My Projects list in the My Projects section of the sidebar, and displays the screen for configuring the slide show. The slide show's first slide (H) and title (I) appear above a bar of thumbnails (J) of the other slides. You can rearrange the slides by dragging them in the thumbnails bar. You can click **Add** (⊕, K) and then **Add Photos** on the pop-up menu to add more photos to the slide show.

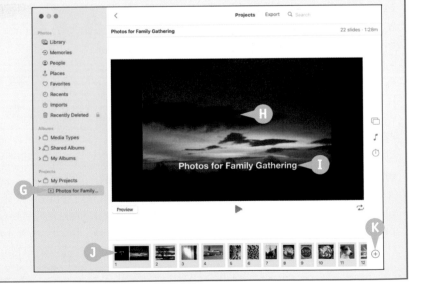

Configure Your Slide Show

After adding all the photos and setting the slide order, configure your slide show. Click **Themes** (⬚, L) to display the Themes pane, and then click the theme you want; if you choose the Ken Burns theme, you then configure the start and end points for panning and zooming. Click **Music** (♪, M) to display the Music pane, and select the music. Click **Duration** (⏱, N), and then select duration settings.

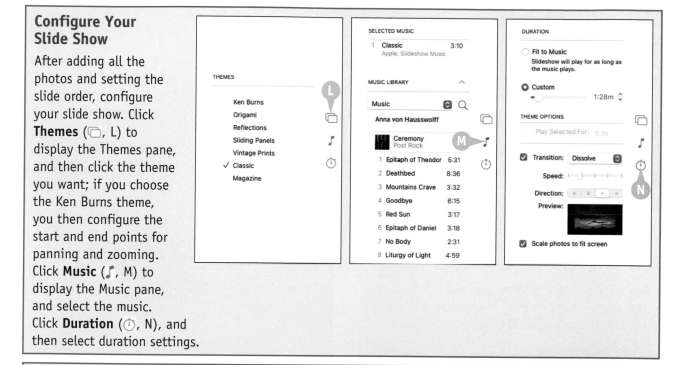

Play Your Saved Slide Show

Once you have configured your slide show, click **Preview** to preview the show, and then make further adjustments, as needed. When the show is ready, click **Play** (▶) to start the show playing. You can then control the slide show by using the pop-up control bar (O); move the pointer to display this bar again after it disappears.

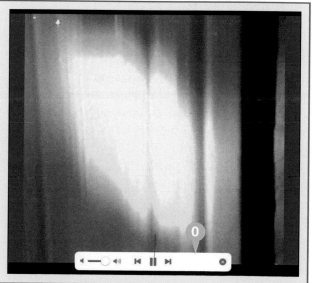

E-Mail a Photo

From Photos, you can quickly create an e-mail message containing one or more photos you want to send to a contact or multiple contacts. You can specify the subject line for the message, add any text needed to explain what you are sending, and choose between including the full version of the photo and creating a smaller version of it that will transfer more quickly.

E-Mail a Photo

1 In Photos, click the photo you want to send via e-mail.

2 Click **Share** (⬆) on the toolbar.

The Share pop-up menu opens.

3 Click **Mail** (✉).

Note: The first time you give the Mail command, you may need to follow through a procedure to set up Photos with your e-mail account.

Photos creates a message containing the photo.

4 Type the recipient's address.

A If Photos displays a pop-up menu of matching contacts, click the correct address.

5 Click **Subject** and type the subject for the message.

6 Click in the message box and type any text needed to explain why you are sending the photo.

7 Click **Image Size** (⬆️).

The Image Size pop-up menu opens.

Ⓑ The tool tip shows the resolution of the photo at this size.

8 Click the size of photo to send — for example, **Large**. See the tip for recommendations.

9 Click **Send** (✈️).

Photos sends the message.

TIP

Which size should I use for sending a photo?
In the Image Size pop-up menu, choose **Small** if the recipient needs only to view the photos at a small size in the message. Choose **Medium** to let the recipient view more detail in the photos in the message. Choose **Large** to send versions of the photos that the recipient can save and use in albums or web pages. Choose **Actual Size** to send the photos unchanged so that the recipient can enjoy, edit, and use them at full resolution.

Shoot Photos or Movies of Yourself

Your MacBook includes a built-in FaceTime HD camera that is great not only for video chats with Messages and FaceTime, but also for shooting photos and movies of yourself using the Photo Booth application. You can use Photo Booth's special effects to enliven the photos or movies. The special effects include distorted views, color changes such as Thermal Camera and X-Ray, and preset backgrounds that you can use to replace your real-world background.

Shoot Photos or Movies of Yourself

1 Click **Launchpad** (⊞) on the Dock.

The Launchpad screen appears.

2 Click **Photo Booth** (⬚).

Note: You can use some macOS-compatible external cameras with Photo Booth. Alternatively, you can use an iPhone connected via the Continuity Camera feature. To change cameras, click **Camera** on the menu bar, and then click the appropriate camera on the menu.

Photo Booth opens.

Ⓐ The photo well shows photos taken earlier in Photo Booth.

3 If your face appears off center, either rotate or tilt your MacBook's screen or move yourself so that your face is correctly positioned.

4 Choose the type of picture to take:

Ⓑ For four pictures, click **Take four quick pictures** (⬛).

Ⓒ For a single still, click **Take a still picture** (▣).

Ⓓ For a movie, click **Record a movie clip** (◲).

5 To add effects to the photo or movie, click **Effects**.

The Photo Booth window shows various effects applied to the preview.

E To see more effects, click **Previous** (◀) or **Next** (▶) or click a different dot in the bar. The center effect on each screen is Normal. Use this effect to remove any other effect.

6 Click the effect you want to use. This example uses **Thermal Camera**.

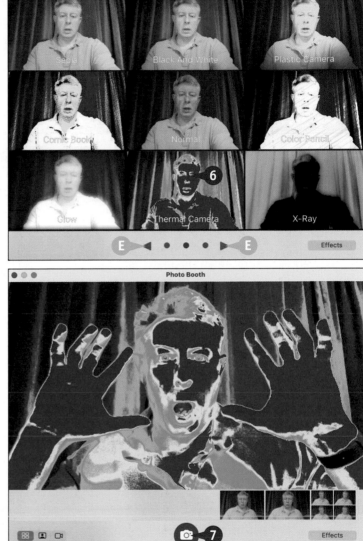

7 Click **Take Photo** (📷) or **Take Movie** (📹).

Photo Booth counts down from 3 and then takes the photo or photos, or starts recording the movie.

8 If you are taking a movie, click **Stop Recording** (⏹) when you are ready to stop (not shown).

Photo Booth adds the photo or movie to the photo well.

TIP

How can I use the photos and movies I shoot in Photo Booth?
After shooting a photo or movie, click it in the photo well, and then click **Share** (⬆). Photo Booth displays a panel with buttons for sharing the photo or movie. Click **Mail** to send it in a message. Click **Add to Photos** to add it to Photos. Click **Change Profile Picture** to use it as your account picture. Click **Messages** to make it your picture in Messages.

CHAPTER 12

Networking, Security, and Troubleshooting

macOS enables you to share files, printers, and scanners across networks. To keep your MacBook running well, you need to use macOS's security features and perform basic maintenance, such as emptying the Trash and updating both macOS and your apps with the latest fixes. You may also need to troubleshoot your MacBook.

Transfer Files Using AirDrop

m acOS's AirDrop feature enables you to transfer files easily via Bluetooth and Wi-Fi between your MacBook and nearby Macs, iPhones, and iPads. Activating AirDrop in a Finder window shows you available Macs, iPhones, and iPads, and you can drag a file to the Mac or device to which you want to send it. Similarly, nearby Macs, iPhones, and iPads can send files to your MacBook via AirDrop, and you can decide whether to accept or reject each file.

Transfer Files Using AirDrop

Send a File via AirDrop

1 Click **Finder** () on the Dock.

A Finder window opens.

Note: If AirDrop does not appear in the sidebar, click **Go** on the menu bar and then click **AirDrop**. If AirDrop does not appear on the Go menu either, your Mac is not compatible with AirDrop.

2 Click **AirDrop** (⊚).

The AirDrop screen appears.

3 Click **Allow me to be discovered by** (⌄).

The pop-up menu opens.

4 Click **No One**, **Contacts Only**, or **Everyone** to specify which people's Macs, iPhones, and iPads can see your MacBook via AirDrop.

5 Press Control +click **Finder** () on the Dock.

The contextual menu opens.

6 Click **New Finder Window**.

A second Finder window opens.

7 Arrange the Finder windows so you can see both.

8 Drag the file to the icon for the Mac, iPhone, or iPad you want to send it to.

The Mac, iPhone, or iPad prompts the user to accept the file.

If the user accepts the file, the Finder sends the file to the recipient.

Note: If the Mac, iPhone, or iPad user declines the file, the Finder displays a dialog telling you so.

Receive a File via AirDrop

When someone sends you a file via AirDrop, a dialog appears on-screen.

A For a picture, the dialog shows a preview.

B You can click **Decline** to decline the file.

1 Click **Accept** (⌄).

The pop-up menu opens.

2 Click the appropriate item. For example, for a photo, you can choose between clicking **Save to Downloads** to save the file to the Downloads folder and clicking **Open in Photos** to add the photo to your photo library.

TIP

Should I use AirDrop or a shared folder on the network to transfer files?

If your MacBook connects to a network with shared folders, use those folders instead of AirDrop. By storing a file in a shared folder, you and your colleagues can work on it directly without transferring copies back and forth. AirDrop is useful for sharing on networks that do not have shared folders or for sharing files with Macs, iPhones, or iPads to which your MacBook does not normally connect. As an alternative, you can use Messages or Mail to transfer files. For security, turn AirDrop off when you are not using it.

Connect to a Shared Folder

macOS enables each Mac to share folders with other computers on the same network. You can connect your MacBook to other Macs and work with the files in their shared folders.

The user who sets up the sharing can assign other users different levels of access to the folder. Depending on the permissions set for the folder, you may be able to view files in the folder but not alter them, or you may be able to not only view but also create, change, and delete files in the folder.

Connect to a Shared Folder

1 Click **Finder** (🙂) on the Dock.

A Finder window opens.

2 If the Locations category is collapsed, position the pointer over it, and then click **Expand** (>) to expand it.

3 Click **Network** (🌐).

The list of devices on the network appears.

4 Click the computer that is sharing the folder.

5 Click **Connect As**.

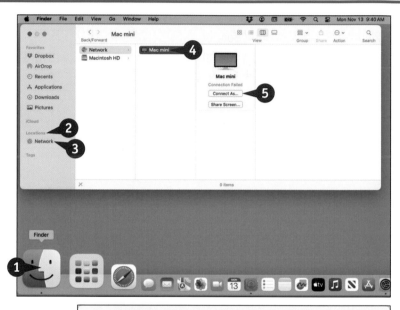

The Connect As dialog opens.

6 Click **Registered User** (○ changes to ●) if you have a user account on the Mac. Otherwise, click **Guest** (○ changes to ●) and go to step **9**.

7 Type your username.

8 Type your password.

Ⓐ You can select **Remember this password in my keychain** (☑) if you want to store your password for future use.

9 Click **Connect**.

The Connect As dialog closes.

B The shared folders appear.

Note: The shared folders you see are the folders you have permission to access. Other users may be able to access different folders.

10 Navigate to the folder whose contents you want to see.

11 Work with files as usual. For example, open a file to work on it, or copy it to your MacBook.

12 When you finish using the shared folder, click **Eject** (⏏) next to the computer's name in the Locations list.

Your MacBook disconnects from the computer sharing the files.

C You can also click **Disconnect** to disconnect from the sharing computer.

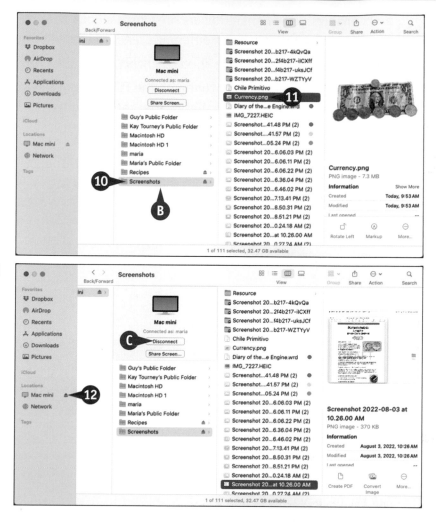

TIP

How can I connect to a shared folder that does not appear in the Shared list in the Finder window?
If the shared folder does not appear in the Shared list, find out the name or IP address of the computer sharing the folder. Click **Go** on the menu bar and click **Connect to Server**. The Connect to Server dialog opens. In the Server Address field, type or paste the computer's name or IP address; for a Windows computer, type **smb://** and then the IP address, such as 192.168.1.55. Click **Connect**, and then provide your username and password if prompted. To reconnect to a server you have used before, click the server in the Favorite Servers list box, and then click **Connect**.

Share a Folder

macOS enables you to share folders on your MacBook with other users on the network. You can set different permissions for different users, such as allowing some users to change files, allowing other users to view files but not change them, and allowing other users no access at all.

To share a folder, you enable and configure the File Sharing service in Sharing settings. System Settings configures sharing for Macs automatically. You can configure sharing for Windows users manually.

Share a Folder

1. Click **System Settings** (⚙) on the Dock.

   The System Settings window opens.

2. Click **General** (⚙).

   The General pane appears.

3. Click **Sharing** (⬦; not shown).

   The Sharing pane appears.

4. Set the **File Sharing** switch to On (⬤).

   System Settings turns on file sharing.

5. Click **Info** (ⓘ).

   The Info dialog for file sharing opens.

6. Click **Add** (+) under the Shared Folders box.

   A dialog for choosing a folder opens.

7. Click the folder you want to share.

8. Click **Open**.

Note: Each user account includes a drop box folder into which other people can place files and folders but whose contents only the user can see. To access this folder, click **Finder** (🙂) on the Dock, click **Go** on the menu bar, and then click **Home**. Click or double-click **Public**, depending on the Finder view, and then click or double-click **Drop Box**.

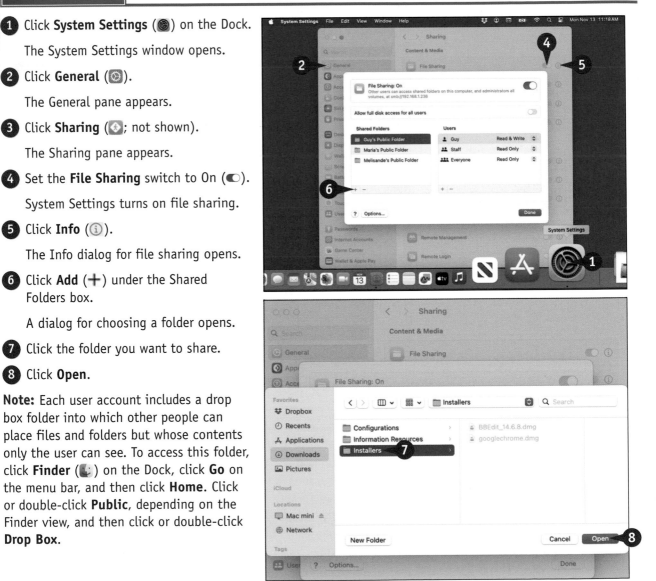

The dialog closes, and the folder appears in the Shared Folders list.

9 Click the folder.

The folder becomes selected.

10 Click **Everyone**.

11 Click the pop-up menu ($\updownarrow$) and select the appropriate permission. See the tip for details.

12 If you need to configure sharing for Windows users, click **Options**. Otherwise, go to step **17**.

The Options dialog opens.

13 Set the **Share files and folders using SMB** switch to On ($\bullet$).

14 Select **On** for the user ($\checkmark$).

15 In the dialog that opens, type the user's password, and then click **OK** (not shown).

16 Click **Done**.

The Options dialog closes.

17 Click **Done**.

The Info dialog for file sharing closes.

18 Click **Close** ($\bullet$).

System Settings closes.

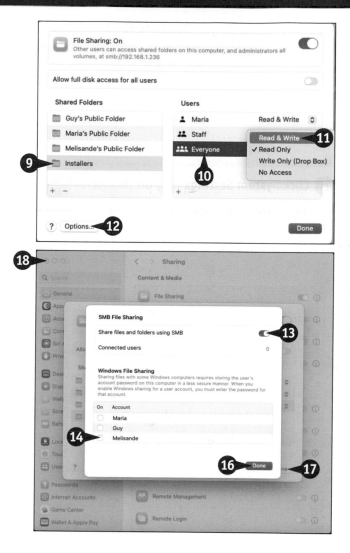

TIP

What permissions should I assign to a folder I share?

Assign **Read Only** permission if you want other people to be able only to open or copy files in the folder. If you want other people to be able to create and change files in the folder, including renaming and deleting them, click **Read & Write**. If you need to create a drop box folder that people cannot view but can add files to, click **Write Only (Drop Box)**.

Connect to a Shared or Network Printer

macOS enables you to connect to shared printers and network printers and print documents to them. By sharing printers, you can not only enable each computer to print different types of documents as needed, but also reduce the costs of printing.

To use a shared printer or network printer, you first set it up on your MacBook using Printers & Scanners settings. After you set up the printer, you can access it from the Print dialog just like a printer connected directly to your MacBook.

Connect to a Shared or Network Printer

1 Click **System Settings** (⚙) on the Dock.

The System Settings window opens.

2 Click **Printers & Scanners** (🖨), toward the bottom of the sidebar.

The Printers & Scanners pane appears.

3 Click **Add Printer, Scanner, or Fax**.

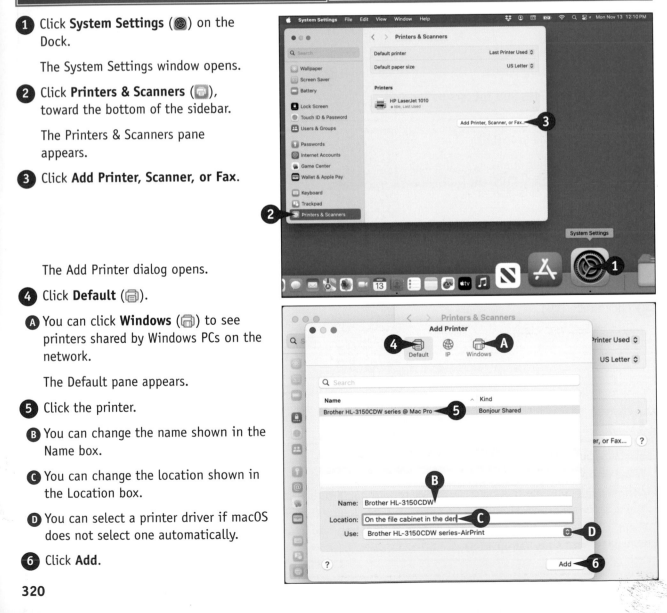

The Add Printer dialog opens.

4 Click **Default** (🖨).

Ⓐ You can click **Windows** (🖨) to see printers shared by Windows PCs on the network.

The Default pane appears.

5 Click the printer.

Ⓑ You can change the name shown in the Name box.

Ⓒ You can change the location shown in the Location box.

Ⓓ You can select a printer driver if macOS does not select one automatically.

6 Click **Add**.

320

A dialog opens showing the progress as macOS downloads and installs any software needed.

The dialog then closes automatically.

The Add Printer dialog closes.

E The printer appears in the Printers & Scanners pane.

7 Click the printer's button.

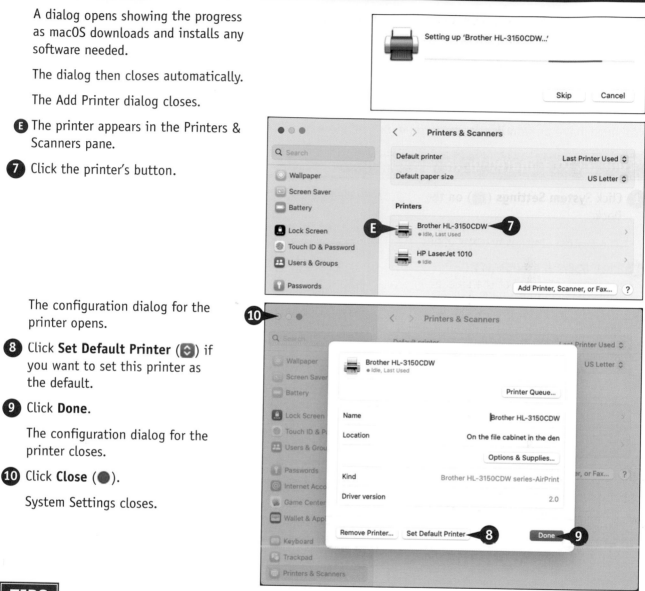

The configuration dialog for the printer opens.

8 Click **Set Default Printer** (◆) if you want to set this printer as the default.

9 Click **Done**.

The configuration dialog for the printer closes.

10 Click **Close** (●).

System Settings closes.

TIPS

What should I do when the Use pop-up menu says "Choose a Driver"?

Click **Use** (◆) and then click **Select Software**. The Printer Software dialog opens. Type a distinctive part of the name in the Search box to see a list of matching items, and then click the driver for the printer model. Click **OK**.

Can I print without connecting to a printer?

You can create a PDF file — an image of the document — that you can print elsewhere. Click **File** and then click **Print** to open the Print dialog. Click **PDF** (✓) and then click **Save as PDF**. In the dialog that opens, specify the filename and location, and then click **Save**.

Turn Off Automatic Login

If your MacBook does not use the FileVault disk-encryption feature, you can set your MacBook to log in one user account automatically, bypassing the login screen. Automatic login is convenient if you are the only person who can access your MacBook, but for security, you should disable automatic login so each user must log in.

To enable or disable automatic login, you work in the Users & Groups pane in the System Settings app. You must have an administrator account or provide administrator credentials to change these options.

Turn Off Automatic Login

1 Click **System Settings** (⚙) on the Dock.

   The System Settings window opens.

2 Click **Users & Groups** (👥).

   The Users & Groups pane appears.

Note: If the Automatically Log In As pop-up menu is set to Off, automatic login is already disabled. Go to step **7**.

3 Click the **Automatically log in as** pop-up menu (⬍).

The pop-up menu opens.

4 Click **Off**.

The Users & Groups dialog opens.

5 Type your password.

6 Click **Unlock**.

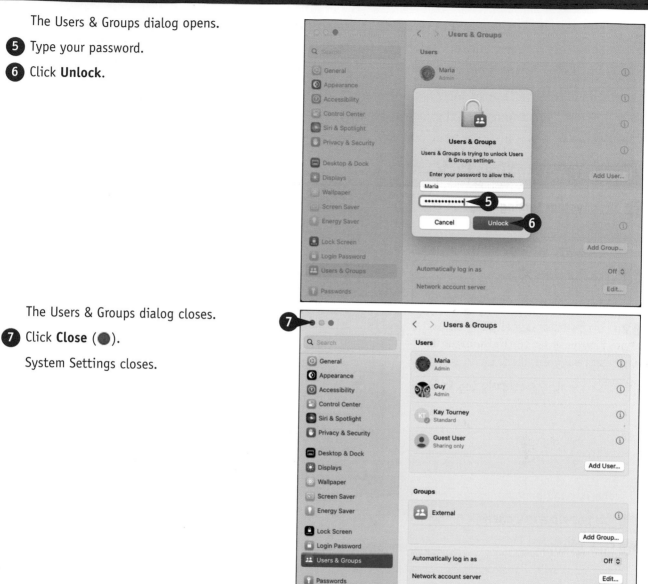

The Users & Groups dialog closes.

7 Click **Close** (●).

System Settings closes.

TIP

What other options can I set to tighten my MacBook's security?
Click **System Settings** (●) on the Dock to open the System Settings window, and then click **Lock Screen** (🔒) to display the Lock Screen pane. In the When Switching User area, click **Name and password** (○ changes to ●) to hide the list of usernames so that anyone logging on must type a username as well as a password. Set the **Show the Sleep, Restart, and Shut Down buttons** switch to Off (⬤) to remove these buttons from the login screen, so that nobody can shut down the MacBook without logging in unless they turn off the MacBook's power. Set the **Show password hints** switch to Off (⬤) if you want to prevent password hints from appearing.

Enable and Configure the Firewall

macOS includes a firewall that protects your MacBook from unauthorized access by other computers on your network or on the Internet. macOS enables you to configure the firewall to suit your needs. To configure the firewall, you use the Firewall pane in Network settings.

Even if your Internet router includes a firewall configured to prevent Internet threats from reaching your network, you should use the macOS firewall to protect against threats from other computers on your network.

Enable and Configure the Firewall

1. Click **System Settings** (⚙) on the Dock.

   The System Settings window opens.

2. Click **Network** (🌐).

   The Network pane appears.

3. Click **Firewall** (🔒).

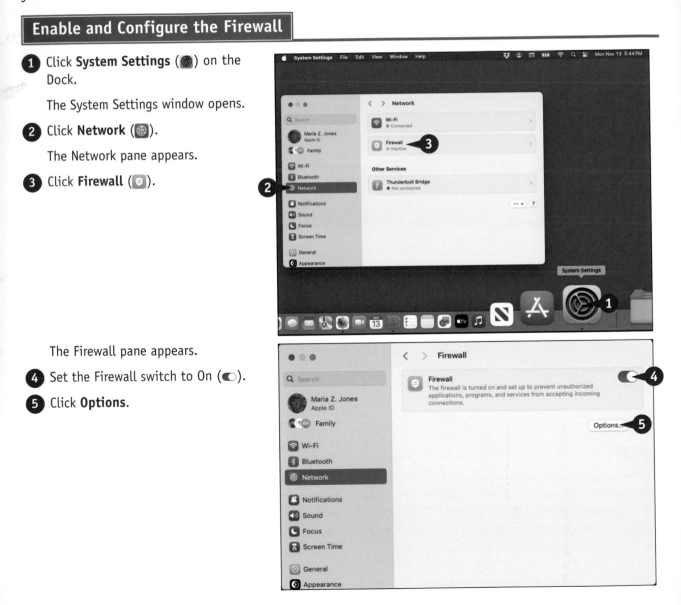

The Firewall pane appears.

4. Set the Firewall switch to On (●).

5. Click **Options**.

The Options dialog opens.

6 Set the **Automatically allow built-in software to receive incoming connections** to On (⚫) if you want to allow your MacBook's built-in software to receive incoming connections.

7 Set the **Automatically allow downloaded signed software to receive incoming connections** switch to Off (◯) if you want to prevent apps you have installed from accepting inbound connections automatically.

8 To allow incoming connections to a particular application, click **Add** (➕).

The Add dialog opens.

9 Click **Applications** (🅰️).

The contents of the Applications folder appear.

10 Click the app.

11 Click **Open**.

The Add dialog closes.

System Settings adds the app to the list.

12 Click **OK**.

The Options dialog closes.

13 Click **Close** (⚫).

System Settings closes.

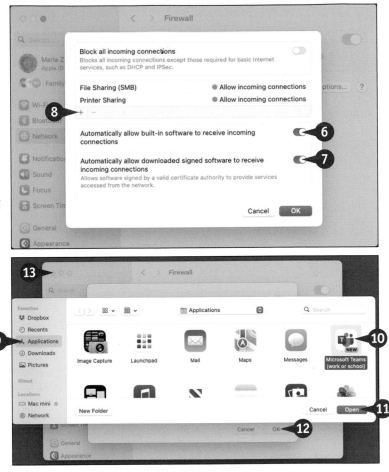

TIPS

When should I enable Block All Incoming Connections?
Set the **Block all incoming connections** switch to On (⚫) when you need to tighten security as much as possible. The usual reason for blocking all connections is connecting your MacBook to a network that you cannot trust, such as a public wireless network or hotspot.

How do I block incoming connections only to a specific application?
Add that application to the list in the Options dialog. Then click the application's **Allow incoming connections** pop-up menu (↕) in the list and click **Block incoming connections**.

Choose Privacy Settings for Location Services

The settings in the Privacy & Security pane in System Settings enable you to control which apps can request access to your MacBook's location. Spend a few minutes to configure Location Services so that it does not provide apps or other services with information you would prefer to keep private.

macOS also enables you to turn off personalized advertisements delivered via the Apple advertising platform. Turning off personalized advertisements does not reduce the volume of ads you receive; it just means the ads will be less relevant to your interests — real or putative — in your advertising profile.

Choose Privacy Settings for Location Services

1 Click **System Settings** (⚙) on the Dock.

The System Settings window opens.

2 Click **Privacy & Security** (🛡).

The Privacy & Security pane appears.

3 Click **Location Services** (➤).

The Location Services pane appears.

4 Set the **Location Services** switch to On (⬤) if you want to allow apps and features to use Location Services. If you prefer not to enable Location Services, set this switch to Off (◯), and go to step **11**.

5 In the list of apps and features that have requested to use Location Services, set each switch to On (⬤) or Off (◯), as needed.

6 Click **Details**.

The Details dialog opens.

7 In the System Services Can Access Your Location For box, set each switch to On (⚪) or Off (⚪), as needed.

8 If you set the Significant Locations switch to On (⚪), click **Details** to open the Significant Locations dialog, inspect the Recent Records list of significant locations, and then click **Done**.

Note: You can clear the Recent Records list by clicking **More** (⋯) and then clicking **Clear History**.

9 Set the **Show location icon in Control Center when System Services request your location** switch to On (⚪) if you want Control Center to display a telltale arrow (➤) when System Services asks for your location.

10 Click **Done**.

The Details dialog closes.

11 Click **Close** (●).

System Services closes.

Reclaim Space by Emptying the Trash

In macOS, the Trash is a receptacle for files you delete from your MacBook's drives. Any file you place in the Trash remains there until you empty the Trash or until the Trash runs out of space for files and automatically deletes the earliest files it contains; until then, you can recover the file from the Trash. You can reclaim drive space by emptying the Trash manually.

Reclaim Space by Emptying the Trash

Empty the Trash

1 Click **Trash** (🗑️) on the Dock.

Note: If the Trash icon on the Dock is the empty Trash can (🗑️), the Trash is already empty.

A Finder window opens showing the contents of the Trash.

Ⓐ You can click **Icon View** (🔲), **List View** (☰), **Column View** (🔳), or **Gallery View** (▭) to change the view, as usual.

2 Look through the files and folders in the Trash to make sure it contains nothing you want to keep.

Note: To quickly view the contents of a file, use Quick Look. Click the file, and then press Spacebar.

Note: You cannot open a file while it is in the Trash. If you want to open a file, you must remove it from the Trash.

3 Click **Empty**.

A dialog opens to confirm that you want to permanently erase the items in the Trash.

Note: You can turn off the confirmation of deleting files. Click **Finder** and then click **Settings**. Click **Advanced** and then deselect **Show warning before emptying the Trash** (⭕).

4 Click **Empty Trash**.

macOS empties the Trash and then closes the Finder window.

Delete a File or Folder Immediately Without Emptying the Trash

1 In the Trash folder, **Control** + click the file or folder.

2 Click **Delete Immediately**.

A confirmation dialog opens.

3 Click **Delete**.

macOS deletes the file or folder without affecting the rest of the Trash.

Restore a File or Folder to Its Previous Location

1 In the Trash folder, press **Control** + click the file or folder.

2 Click **Put Back**.

Note: You can also click the item, click **Action** (⊙ ∨) to open the Action menu, and then click **Put Back**.

macOS restores the file or folder to its previous location.

Note: To move a file from the Trash to another folder, drag the file to that folder. For example, you can drag a file to the desktop.

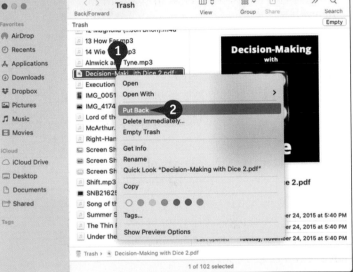

TIPS

Is there a quicker way to empty the Trash?

If you are sure that the Trash contains no files or folders you need, press **Control** + click **Trash** (🗑) on the Dock. The contextual menu opens. Click **Empty Trash**. A confirmation dialog opens. Click **Empty Trash**. Alternatively, press ⌘ + **Shift** + **Del**.

Why does the Put Back command not appear on the Action menu or the contextual menu?

If the Put Back command is missing, the folder from which you moved the item to the Trash is no longer where it was. If the folder is in the Trash, click the folder, click **Action** (⊙ ∨), and then click **Put Back**; you can then put back into the folder the file you want to recover. Alternatively, drag the file to another folder.

Keep Your MacBook Current with Updates

You can use the App Store app built into macOS to check for updates to your MacBook, its operating system, and your App Store apps. When the App Store app finds updates, you can choose which to install.

Your MacBook must be connected to the Internet when you check for updates and download them. You can install most updates when your MacBook is either online or offline. Some updates require restarting your MacBook.

Keep Your MacBook Current with Updates

Note: macOS may display an Updates Available dialog prompting you to install updates. You can click **Updates Available** to go to the Updates tab in the App Store app, click **Install** to install the updates immediately, or click **Later** to install the updates later.

(A) A badge on the App Store icon indicates the number of updates available.

(1) Click **App Store** (⬛) on the Dock.

Note: You can also open the App Store app by clicking **Apple** (🍎) and then clicking **App Store**.

The App Store app opens.

(2) Click **Updates** (⬇).

The Updates pane appears.

App Store automatically checks for updates.

If updates are available, the Available list shows the details.

Note: If the message No Updates Available appears, go to step **4**.

(B) You can install an individual update by clicking **Update**. If the update requires a restart, click **Restart**.

(C) You can click **more** to display full details of an update.

(3) To install all the updates, click **Update All**.

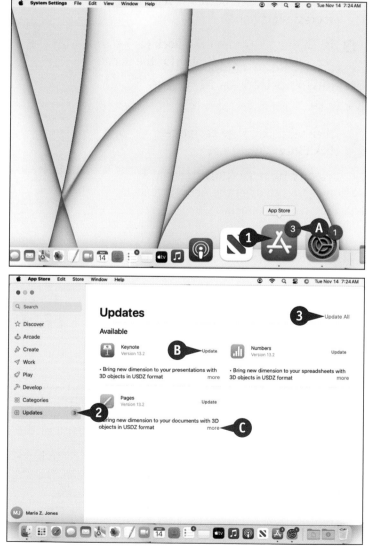

Note: If installing the updates requires a restart, App Store displays a dialog telling you so. Click **Restart** to restart now. Alternatively, click **Later** and then click **Install in an Hour**, **Install Tonight**, or **Remind Me Tomorrow**.

macOS installs the updates.

D A progress indicator appears for each app.

E When the app update is complete, App Store moves the app into the Updated Recently category. The Open button replaces the progress indicator.

F The Launchpad icon on the Dock ( ) shows an overall progress indicator for the updates.

Note: macOS restarts your MacBook if necessary.

4 When all updates are complete, click **Close** ( ).

The App Store app closes.

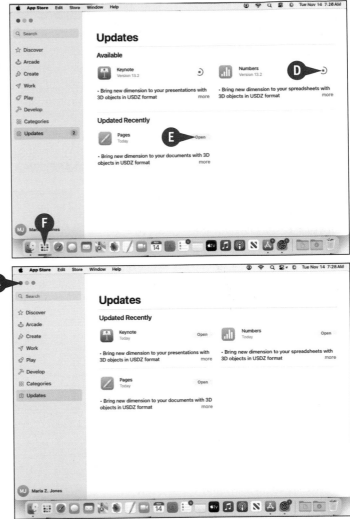

TIPS

Why does the App Store app sometimes prompt me to install updates?

In macOS, App Store comes set to check for updates automatically; when it finds updates, it prompts you to install them. You can change the frequency of these checks, choose whether to download important updates automatically, or turn off automatic checks. See the next section, "Control Checking for Software and App Updates," for instructions.

Which updates should I install?

Install all available system updates unless you learn that a specific update may cause problems with your MacBook. In that case, wait until Apple fixes the update. For app updates, you may prefer to wait for user feedback, because some updates create incompatibilities for documents created in earlier versions.

Control Checking for Software and App Updates

In macOS, the Software Update feature handles updates for the operating system and built-in apps, while the App Store app handles updates for apps you add via the App Store. To keep your MacBook running smoothly and protect it from both online and offline threats, you should apply system software updates when they become available. You may also want to install app updates soon to take advantage of any bug fixes or improvements they offer.

Control Checking for Software and App Updates

Set Software Update to Update Your MacBook Automatically

A A badge indicates a software update is available.

1 Click **System Settings** (⚙) on the Dock to open the System Settings window.

2 Click **General** (⚙) in the sidebar to display the General pane.

3 Click **Software Update** to display the Software Update pane.

B If an update is available, you can click **Update Now** when you are ready to install the update.

4 Click **Info** (ⓘ) to open the Info dialog.

5 Set the **Check for updates** switch, the **Download new updates when available** switch, and the **Install macOS updates** switch to On (⬤).

6 Set the **Install application updates from the App Store** switch to On (⬤) if you want to install app updates.

7 Set the **Install Security Responses and system files** switch to On (⬤).

8 Click **Done** to close the Info dialog.

9 Click **Close** (⬤) to close System Settings.

Set the App Store App to Install Updates Automatically

1 Click **App Store** (⬜) on the Dock to open the App Store app.

2 Click **App Store**.

3 Click **Settings** on the menu.

The Settings window opens.

4 Select **Automatic Updates** (☑) to have App Store download and install app updates automatically.

5 Select **Automatically download apps purchased on other devices** (☑) to automatically install apps you buy or get on your other Apple devices.

6 Select **Automatically download in-app content** (☑) to have apps download content before you launch them.

7 Select **Video Autoplay** (☑) to have App Store automatically play preview videos with the sound off.

8 Select **In-App Ratings & Reviews** (☑) if you want to allow apps to prompt you for ratings and reviews.

9 Click **Close** (⬤).

The Settings window closes.

10 Click **Close** (⬤).

The App Store app closes.

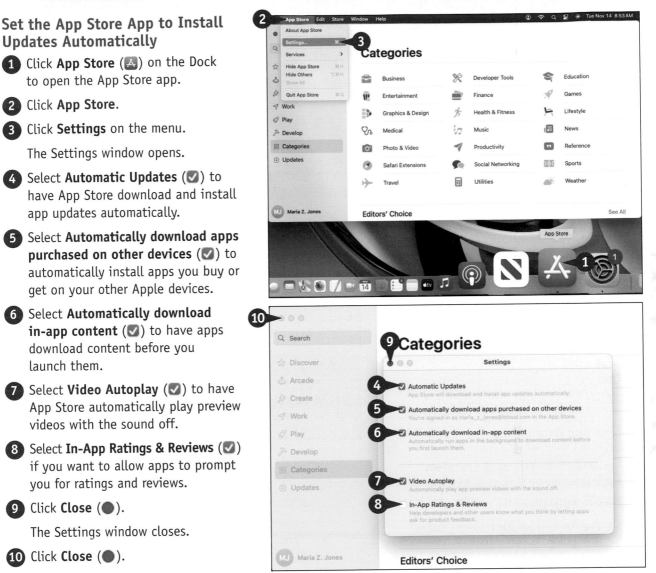

TIP

Should I set the Install macOS Updates switch in the Info dialog to On (⬤)?

Yes. Almost all security professionals recommend installing every update to keep your MacBook protected against online malefactors and to make sure you have the latest fixes and upgrades for macOS and your apps.

If you choose not to install all updates, be sure to install the Security Responses and system files. These are critical to keeping your MacBook protected against the latest threats and attacks.

Back Up Your Files

To enable you to keep your valuable files safe, macOS includes an automatic backup application called Time Machine. Time Machine automatically saves copies of your files to an external drive or a shared folder on your network. You can choose what drive to use, how frequently to back up your files, and what folders to include.

To protect your data, you must back up your files. Time Machine is the most convenient choice because it takes only a few minutes to set up and thereafter runs automatically.

Back Up Your Files

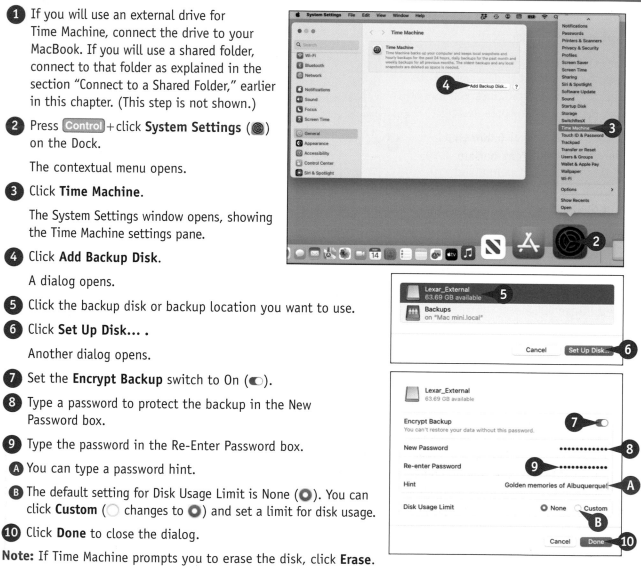

1 If you will use an external drive for Time Machine, connect the drive to your MacBook. If you will use a shared folder, connect to that folder as explained in the section "Connect to a Shared Folder," earlier in this chapter. (This step is not shown.)

2 Press Control + click **System Settings** (⚙) on the Dock.

The contextual menu opens.

3 Click **Time Machine**.

The System Settings window opens, showing the Time Machine settings pane.

4 Click **Add Backup Disk**.

A dialog opens.

5 Click the backup disk or backup location you want to use.

6 Click **Set Up Disk... .**

Another dialog opens.

7 Set the **Encrypt Backup** switch to On (⬤).

8 Type a password to protect the backup in the New Password box.

9 Type the password in the Re-Enter Password box.

Ⓐ You can type a password hint.

Ⓑ The default setting for Disk Usage Limit is None (⬤). You can click **Custom** (◯ changes to ⬤) and set a limit for disk usage.

🔟 Click **Done** to close the dialog.

Note: If Time Machine prompts you to erase the disk, click **Erase**.

334

C The disk appears in the Time Machine pane.

11 Click **Options** to open the Options dialog.

12 Click **Backup Frequency** (⟨⟩) and click the frequency you want. See the second tip for advice.

13 Set the **Back up on battery power** switch to On (⬤) only if you want your MacBook to perform backups even when it is running on battery power. Backing up on battery power will reduce the MacBook's runtime on the battery.

14 Click **Add** (✚).

A dialog opens.

15 Select each drive or folder you want to exclude from backup.

16 Click **Exclude**.

The dialog closes, and Time Machine adds the items to the Exclude These Items from Backups dialog.

17 Click **Done** to close the Options dialog (not shown).

18 Click **Close** (⬤).

System Settings closes.

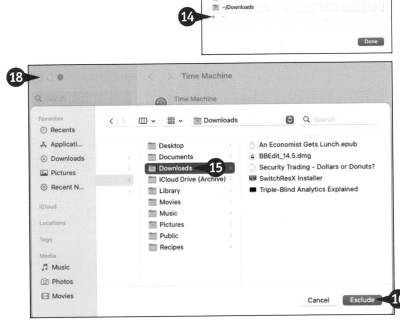

TIPS

What kind of drive should I use for Time Machine?
An external USB drive is usually the best bet, because many such drives are available; they deliver good performance and they are usually affordable. A Thunderbolt drive can deliver higher performance but typically costs much more. Buy a drive that has at least twice as much storage as your MacBook has, and preferably much more, to ensure that you have plenty of space for backups.

Which backup frequency should I choose?
Your choices are Automatically Every Hour, Automatically Every Day, Automatically Every Week, and Manually. By far the best choice is Automatically Every Hour. You can also run extra backups manually for extra protection.

Recover Files from Backup

The Time Machine feature enables you to easily recover files from your backups. So when you delete a file by accident or discover that a file has become corrupted, you can recover the file from backup by opening Time Machine. You can recover either the latest copy of the file or an earlier copy.

If you still have the current copy of the file that you recover, you can choose whether to overwrite that copy or keep it. Time Machine refers to this copy as the "original" file.

Recover Files from Backup

Note: This section shows you how to recover files using the Finder. To recover old data within Contacts, Calendar, or Mail, open the appropriate app and make it active before giving the Enter Time Machine command.

1 Click **Finder** (🙂) on the Dock.

A Finder window opens.

2 Navigate to the folder that used to contain the file or folder you want to recover.

3 Click **Time Machine** (⏲) on the menu bar.

The Time Machine menu opens.

4 Click **Browse Time Machine Backups**.

Note: If Time Machine (⏲) does not appear on the menu bar, click **Launchpad** (▦) on the Dock, and then click **Time Machine** (◉) on the Launchpad screen.

Time Machine opens.

Ⓐ The front window shows your MacBook's drive or drives in their current state.

Ⓑ Backups of the selected drive or folder appear in the windows behind it, with the newest at the front.

Ⓒ The timeline on the right shows how far back in time the available backups go.

5 Click the date or time from which you want to recover the files or folders.

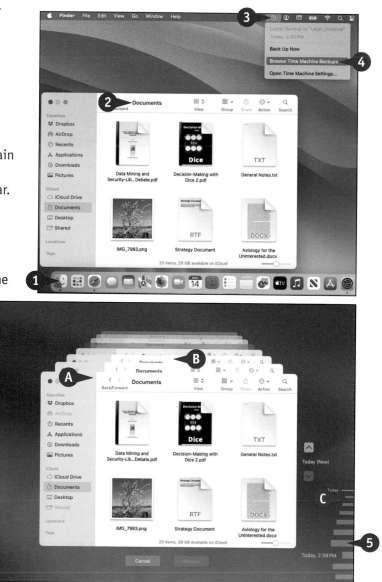

Time Machine brings the backup you chose to the front.

6 Select the item or items you want to restore.

7 Click **Restore**.

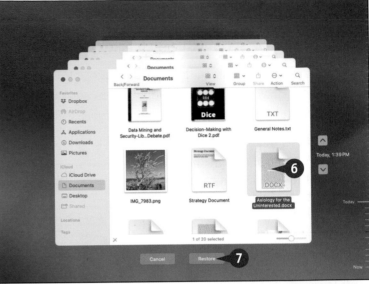

Time Machine disappears.

If restoring a file will overwrite the current version, the Copy dialog opens.

8 Choose how to handle the file conflict (not shown):

D Click **Replace** to replace the current file with the older file.

E Click **Keep Original** to keep the current file.

F Click **Keep Both** to keep both versions of the file. Time Machine adds "(Original)" to the name of the current version.

TIPS

What do the arrow buttons to the right of the Finder window in Time Machine do?

The two arrow buttons enable you to navigate among the available backups. Click the upward arrow to move to the previous backup, further in the past. Click the downward arrow to move to the next backup, nearer to the present.

How do I create Time Machine backups manually?

Click **Time Machine** (⊙) on the menu bar and select **Back Up Now.** The menu also enables you to see when the most recent backup ran, browse Time Machine backups, and open Time Machine settings.

Recover When macOS Crashes

ormally, macOS runs stably and smoothly, but sometimes the operating system may suffer a crash. Crashes can occur for various reasons, including power fluctuations, bad memory modules, an app that has been corrupted, or problems with disk permissions.

Your MacBook may detect that the crash has occurred and display an informational message, but in other cases the MacBook's screen may simply freeze and continue displaying the same information. Normally, you can recover from a crash by turning off your MacBook's power and then turning it on again.

Recover When macOS Crashes

Recover from the Screen Freezing

1 If the pointer shows the "wait" cursor that looks like a spinning beach ball, wait a couple of minutes to see if macOS can recover from the problem. If the pointer has disappeared, go to step **2**.

2 To verify that your MacBook is not responding, press keys on the keyboard or move your fingers on the trackpad (not shown).

Note: If possible, connect your MacBook to power when attempting to recover from a freeze or a crash.

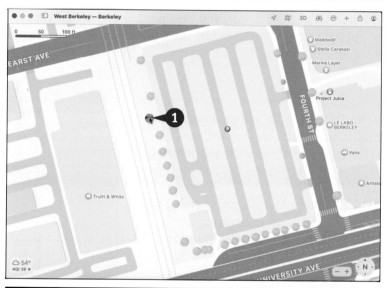

3 Press and hold ⌘+Control and press the MacBook's power button (not shown).

4 If the MacBook does not respond to that key combination, press and hold the MacBook's power button for about 4 seconds (not shown).

The MacBook turns off.

5 Wait at least 8 seconds, and then press the power button once to restart your MacBook (not shown).

Recover from a Detected Crash

When your MacBook detects a macOS crash, it dims the screen and displays a message in the center.

1 Read the message for information.

2 Press and hold the MacBook's power button for about 4 seconds (not shown).

The MacBook turns off.

3 Wait at least 8 seconds, and then press the power button once to restart the MacBook (not shown).

Note: Depending on how your MacBook is configured, macOS may log you in automatically or display the Login dialog.

The login screen appears.

4 Click your username, and then log in to your account as normal.

TIP

How can I avoid crashes?

- Limit the number of apps you run at the same time. When you finish using an app, quit it.
- Keep your MacBook current by installing available updates for system software and apps.
- Keep at least 15 percent of your MacBook's drive free. Click the desktop, click **Go** on the menu bar, and then click **Computer** to open a Finder window showing the Computer folder. Press Control+click **Macintosh HD** and click **Get Info**. In the General section, look at the Capacity readout and the Available readout.
- If running a particular app causes your MacBook to crash, uninstall and reinstall that app.

Recover, Restore, or Reinstall macOS

If your MacBook is suffering frequent freezes or crashes, you should turn to Apple's Recovery toolkit, which enables you to repair disk problems, restore from a backup, or reinstall macOS from scratch. To use Recovery, you start your MacBook from its recovery partition and then use the Recovery utilities. You can then run First Aid from outside macOS, restore your macOS installation from a Time Machine backup, or reinstall macOS from scratch.

Launch macOS Recovery

To launch macOS Recovery on a MacBook with an Apple Silicon processor, such as an M1, M2 Pro, or M3 Max, shut the MacBook down as usual. Then press the Power button and keep holding it down until you see the message *Loading startup options* appear on the screen. Release the Power button and wait for the opening screen to appear. Here, click **Options**, and then click **Continue**. The initial macOS Recovery screen then appears. Click your user name, click **Next**, and authenticate yourself with your password. The Recovery screen then appears.

On a MacBook with an Intel processor, click **Apple** (), click **Restart**, and then click **Restart** in the confirmation dialog that opens. When the startup chime plays, hold down ⌘+R, and keep holding down the keys until the Apple logo appears on the screen. Release the keys and wait for the Recovery screen to appear.

Run First Aid from Recovery Utilities

After launching Recovery Utilities, you can open Disk Utility and run First Aid on a disk.

Click **Disk Utility** () on the Recovery Utilities screen, and then click **Continue**. On the Disk Utility screen, click **Macintosh HD volumes** (A) in the left pane, and then click **First Aid** () on the toolbar.

When First Aid finishes, click **Disk Utility** on the menu bar, and then click **Quit Disk Utility**. The Recovery screen then appears again.

Restore macOS from a Time Machine Backup

If you have been using Time Machine to back up your MacBook's files, you can restore your MacBook from a recent backup.

Click **Restore from Time Machine** on the Recovery Utilities screen, and then click **Continue**. On the Restore from Time Machine screen, click **Continue** again, and then follow the instructions for selecting the backup to use and restoring it to the MacBook.

Time Machine System Restore Edit Window

Time Machine System Restore

Restore from Time Machine

To restore your system from a Time Machine backup or local snapshot, click Continue and follow the instructions.

Important information about restoring your system:
- Restoring will erase the selected destination disk.
- Use Migration Assistant to restore to a new computer.
- To restore individual files use Time Machine in the Finder.

Go Back   Continue

Reinstall macOS from Recovery Utilities

If nothing else has worked, you may need to reinstall macOS on your MacBook. Click **Reinstall macOS Sonoma** on the Recovery Utilities screen to start the process. On the macOS Sonoma screen, click **Continue**, and then follow the prompts.

macOS Sonoma

To set up the installation of macOS Sonoma, click Continue.

Continue

Index